Never Again?

Never Again?

THE THREAT OF

THE NEW

ANTI-SEMITISM

A B R A H A M   H .   F O X M A N

 HarperSanFrancisco
A Division of HarperCollinsPublishers

HarperCollins books may be purchased for educational, business, or sales promotional use. For information please write: Special Markets Department, HarperCollins Publishers, Inc., 10 East 53rd Street, New York, NY 10022.

HarperCollins Web site: http://www.harpercollins.com

HarperCollins®, 📖 ®, and HarperSanFrancisco™ are trademarks of HarperCollins Publishers, Inc.

FIRST HARPERCOLLINS PAPERBACK EDITION PUBLISHED IN 2004

Designed by Jessica Shatan

Library of Congress Cataloging-in-Publication Data has been ordered and is available upon request.

ISBN 0–06–073069–2 (pbk.)

04 05 06 07 08 ❖/RRD 10 9 8 7 6 5 4 3 2 1

To my beloved wife, Golda, my dear children, Michelle and Ariel, and son-in-law, Dan, and my extraordinary grandchildren, Leila and Gideon Small.

To the memory of my dear parents, Helen and Joseph, who survived the Holocaust while members of their families perished along with the six million Jews—victims of anti-Semitism.

CONTENTS

"Never again?" The question mark is Abraham Foxman's, the Anti-Defamation League leader known the world over for his commitment to Israel and the hope it embodies, as well as for his courageous and tireless fight against anti-Semitism, which, in its poisonous hatred, still transcends national frontiers and social structures.

Thus it is a challenge that Foxman delivers to a century that has hardly begun; how, he seems to ask, is it possible for anti-Semitism to resurface undiminished? Have people forgotten its nefarious consequences? Was Hannah Arendt right when she predicted that of the twentieth century's social diseases, anti-Semitism alone would survive, reaching beyond time and geography, religious beliefs, and political affiliations?

Question: What is it about anti-Semitism that keeps it alive? Fascism has been defeated, nazism beaten, communism discredited: Why then is the oldest collective hate-obsession in recorded history immune to change? What makes it so popular, so attractive, so seductive?

The reasons invoked by those who hate Jews, disregarding all truth and logic, combine all possible contradictions. For some, Jews were or are too wealthy; for others, they are or were too poor. Too religious or not enough. Too Jewish or too assimilated. Too learned or too ignorant. Too smart or too naïve. Too nationalistic or too universalist. Certain intellectuals allowed their talent to be tainted by anti-Semitism. Wrtiers such as Ezra Pound, Louis Ferdinand Celine and Knut Hamson Kant and Voltaire,

Fichter and Schopenhauer all made derogatory remarks about Jews. Hitler believed that all Jews were communists; Stalin was convinced that all Jews were capitalists. Hitler and Stalin were mortal enemies; yet they were united in their hatred of Jews. Thus anti-Semitism has a wide range of components: religious and social, ethnic and professional, racist and political. No wonder that a French thinker called it "the socialism of the imbeciles."

In this book, written with understandable urgency, Foxman concentrates on the peril of anti-Semitism today. Having devoted most of his adult life to the defense of Jews and other minorities, often against the same slanderers and practitioners of violence, Foxman has impeccable credentials. The facts he quotes are undisputed, Jewish cemeteries are being desecrated in Europe, synagogues are being destroyed, incitement against Jews does exist, anti-Zionism is often so intense that it borders on hatred, and Holocaust denial is widespread and vicious, as is the official policy in several Arab lands denying the Jewish state's right to exist.

Hence the importance of Foxman's timely book. It is not only a political essay; it is also a personal memoir that will find its place in the literature of memory. His meetings with church leaders, his support of ecumenism, his thoughts on Jewish-Black relations and on various Holocaust-related issues are important if, at times, debatable, in particular his view of fictionalized movies in Holocaust education. The pages describing his survival as a "hidden child" during the Holocaust are superb. Sheltered by a Christian woman who eventually baptized him, he considered her his true mother for a while, even after liberation. But then . . .

Well, read the book. You will not regret it.

Read it and you will learn what it is possible to do with one's memories of a painful past.

Elie Wiesel

ACKNOWLEDGMENTS

Many writers acknowledge a mentor who inspired them to pursue a career or a mission. I have been fortunate to have had many. First and foremost was my father, who taught me Yiddishkeit, history, and instilled in me a sense of pride in being Jewish. He also taught me an appreciation for the lessons of history and about the perils facing the Jewish people.

In my professional life at the Anti-Defamation League (ADL), which began in 1965, upon graduation from NYU Law School in 1965, I entered the ranks of an organization of dedicated professional and lay leaders unite in their determination to lessen the dangers of hate, prejudice, racism, and anti-Semitism. I am indebted to them for their mentoring support and professionalism in the daily fight against the consequences of bigotry. National chairmen I have learned from and had the privilege of working with: Seymour Graubard, Burton M. Joseph, Maxwell E. Greenberg, Kenneth J. Bialkin, Burton S. Levinson, Melvin Salberg, David H. Strassler, Howard P. Berkowitz, and Glen A. Tobias—gratitude for their faith in our mission and in me.

A special source of wisdom and professionalism came from my first boss and colleague, Arnold Forster, who hired me; the late Benjamin R. Epstein, who promoted me; and the late Nathan Perlmutter, who was my partner and teacher at ADL for many years.

I am more than grateful to Karl Weber for having patience with me in the process of working on this book. We became partners in my first literary venture and also became friends. His understanding of what I was thinking and saying made us not only partners, but friends in this venture. His hard work and intelligence truly helped this book come together.

My agent, Lynne Rabinoff—I thank her for her enthusiasm, faith, guidance, support, and dedication in bringing this project to fruition. To my editor, Gideon Weil, I express my profound appreciation for his professional and devoted attention to this literary project, and for his constant support. Thank you to Terri Leonard, senior managing editor, for putting it all together. Special thanks to Roger Freet, associate marketing and publicity director for his hard work in getting behind this project.

Thank you to the ADL staff that assisted me in this effort who are too numerous to mention individually. Three, however, do stand out for special tribute: Kenneth Jacobson, for his wise counsel and advice; Gail Gans, for overseeing the research; and Myrna Shinbaum, for her tireless efforts in helping to keep this project on track and seeing it through to its completion.

Abraham H. Foxman

Never Again?

The challenge with writing a book on anti-Semitism in the world today is that the landscape is constantly changing. In the past year alone, since the hardcover came out, there have been both bad news and good news that warrant notice.

The bad news is that anti-Semitism does not lie dormant for long nor does it take refuge in dark, quiet places. On October 16, 2003, at the Tenth Islamic Summit Conference, the Prime Minister of Malaysia, Mahathir Mohamad, proclaimed that Jews have "gained control of the most powerful countries and they, this tiny community, have become a world power." This was the first anti-Semitic speech made by a head of state since Hitler.

He told the assembled Islamic counties, "The Europeans killed six million Jews out of twelve million. But today the Jews rule this world by proxy. The Muslims will forever be oppressed and dominated by the Europeans and the Jews. We are actually very strong. Some 1.3 billion people cannot be simply wiped out. They get others to fight and die for them. Over a billion Muslims cannot be defeated by a few million Jews. There must be a way." Like Hitler, Mahathir received a standing ovation and thunderous applause for what amounted to a call for a holy war against Jews.

Shocking as this was, even more distressing was the deafening silence from the international community in reaction to the speech, except for the United States. Where was the outcry from world leaders? Where were the voices of responsible leaders of the civilized world saying that you can no longer excuse or rationalize this kind of incendiary rhetoric, this kind of hatred and scapegoating, this blaming of Jews for the ills of the Muslim world? In time and after too much deliberation good people spoke out. The European Union, Italy, Spain and Germany made important comments and efforts to rightly denounce and condemn the speech as anti-Semitic, dangerous, and morally repugnant.

The drumbeat of Jew-hatred emanating from the Arab/Muslim world has not let up. Anti-Semitism continues to permeate its media and culture. Television programs widely available to viewers across the Muslim and Arab world and around the globe portray Jews in ugly and incendiary stereotypes. Here is one example: *Ash-Shatat* (The Diaspora) was a vicious anti-Semitic television series depicting stereotypical Jews hatching a plot for Jewish world control and domination. A Syrian production company aired this in October and November 2003 on the Lebanon-based satellite television network Al-Manar, which is owned by the terrorist organization Hezbollah. Its release was timed to coincide with the Muslim holy month of Ramadan. The thirty-part series purports to dramatize the true history of the rise of modern Zionism and the establishment of the State of Israel, and depicts historical figures, such as Theodor Herzl, Alfred Dreyfus, and others.

The extremely hostile depiction of Jews and the propagation of age-old anti-Semitic conspiracy theories in media, books, film, television, and song is a frightening demonstration of the entrenched anti-Semitism in much of the Arab/Muslim world. Equally as frightening is how far-reaching it is, in this age of the Internet and satellite technology.

In the United States, the number of anti-Semitic incidents remains at a constant and disturbing level. Individuals continue to express their hatred of Jews through acts of violence, vandalism, harassment, and intimidation. For each act against a single person or property, a community suffers. The most violent act in 2003 was the burning down of a Holocaust museum in Terre

Haute, Indiana. A synagogue in Wildwood, New Jersey, had a bullet fired through its front door; in Allentown, Pennsylvania, it was a Molotov cocktail. "Heil Hitler," "Death to Jews" and swastika graffiti appear profusely on public property, as well as on Jewish-owned property.

The nation's largest neo-Nazi group, the National Alliance, in an attempt to garner new support, membership, and publicity for its cause, has blanketed communities around the country with racist, anti-Semitic and anti-gay fliers. And then there is the Internet. Anti-Semitic hate sites operate openly and freely, often targeting the youngest and most impressionable, our children.

Hate crimes continue to occur around the globe, but more often than not these stories do not appear in your local newspaper.

- Colmar, France (April 29/30, 2004)—A Jewish cemetery in the Alsace region dating back to the 18th century was vandalized; at least 127 headstones were spray painted with swastikas and anti-Semitic statements.

- Montreal, Canada (April 4, 2004)—A Jewish school was set on fire by an arsonist, heavily damaging the library of the United Talmud Torah School. Police found a note with anti-Semitic comments on the exterior wall of the library.

- Toulon, France (March 23, 2004)—A Jewish synagogue and community center were set on fire.

- Toronto, (March 19-21, 2004)—A weekend-long spate of anti-Semitic vandalism targeted a Jewish cemetery, a Jewish school, and several synagogues.

- St. Petersburg, Russia (February 15, 2004)—Fifty graves were desecrated in a Jewish cemetery, with swastikas and anti-Semitic graffiti painted on headstones.

- Hinterbruehl, Austria (January 5, 2004)—Vandals burned the words "Kill the Jews" into the lawns of Tasmania's Parliament House.

While anti-Semitism still remains a serious problem in Europe, the good news is that it is being acknowledged and confronted by European leaders who are no longer in denial and are seeking ways to counter it. Two years ago, as violent acts of anti-Semitism were sweeping Europe, leaders were reluctant to acknowledge and confront the issue head-on. Jews around the world sounded the alert and called on the EU, government leaders, and good people in each country to recognize that they had an anti-Semitism problem, and to match words with deeds.

Many did. French President Jacques Chirac finally stopped denying that anti-Semitism existed in France, declared it unacceptable, and created educational and legislative initiatives to combat it. The OSCE Conference in Vienna in June 2003 featured anti-Semitism and called for convening the Berlin Conference. Under the leadership of Romano Prodi of Italy, the EU held a conference in Brussels in February 2004 to discuss the serious implications of the results from an EU poll that showed more Europeans viewed Israel as more of a threat to world peace than any other country. In March 2004, in response to the situation, the EU Monitoring group issued a full report on anti-Semitism in Europe.

Beyond the end of denial, steps have been taken and structures put in place to combat the hatred, most notably in France. President Chirac established a special task force, led by Prime Minister Jean-Pierre Raffarin and including the Ministers of Justice, Interior, Education, and Foreign Affairs.

Already a number of initiatives have emerged from this structure, among them Holocaust education projects for schools and legislation that would prohibit hateful satellite broadcasts from the Middle East. Aside from these substantive efforts, the message emanating from the highest levels of government has been important in itself.

A recent opinion survey conducted for ADL of adults in ten European countries found some decrease in anti-Semitic attitudes from its 2002 findings. It was released in Berlin on the eve of an historic international conference on anti-Semitism, convened by the Organization for Security and Cooperation in Europe (OSCE).

The 2004 results of those who hold anti-Semitic attitudes in the first group of five countries are as follows:

France—25%, down from 35% in 2002

Germany—36%, down from 37%

Belgium—35%, down from 39%

Denmark—16%, down from 21%

U.K—24%, up from 18%

This data, also from 2004, represents the second group of respondents who hold anti-Semitic views:

Spain—24%, down from 34 percent in 2002

Italy—15%, down from 23%

Switzerland—17%, down from 22%

Austria—17%, down from 19%

The Netherlands—9%, up from 7%

Actions bring results, and these findings demonstrate that when denial is out and action is in, not just the number of anti-Semitic acts decrease, but so do many negative attitudes about Jews.

Nevertheless, large numbers of Europeans still accept a wide range of traditional anti-Semitic stereotypes about Jews. Anti-Israel sentiment and beliefs, such as the charge that Jews are more loyal to Israel than to their home countries, help incite and legitimize anti-Semitism.

In the city that once served as the capital of Nazi Germany, fifty-five nations gathered on April 28-29, 2004, for an historic Conference on anti-Semitism convened by the Organization for Security and Cooperation in Europe. As a public advisor to the U.S. Delegation I had the opportunity to address the meeting. The following is the crux of what I offered to the conference:

There are three fundamental truths about bigotry that apply to the current problem of anti-Semitism in the OSCE region. One is that, when beginning to confront a new form of bigotry, people fear that by talking about the problem honestly, they create it. Second is that, until we have a common language and understanding about what the problem is, we cannot come together to combat it. We cannot even monitor it effectively and we certainly cannot seek out solutions. There is no common language—no common definition—no agreement as to what is indeed an act of anti-Semitism. Further, there exists no formal system through which to channel information—if you ask the man or woman on the street to whom they should report anti-Semitism, you will often hear conflicting answers.

Third, confronting and recognizing bigotry honestly often runs against a prevailing political climate. Just as openly confronting bigotry against African Americans in the American South was an irritant in the climate of the day, so today we are struggling to achieve a recognition of the current manifestation of anti-Semitism that is causing the most problems.

Addressing the new forms of anti-Semitism honestly is considered controversial. In the United Nations and even in the OSCE, language on anti-Semitism is not dealt with by the human rights departments, but in the Middle East section. The discussion of anti-Semitism in the context of the rights of Jewish people to have their nationalism, their self-determination, their homeland, is a political hot-button issue.

Anti-Semitism is not a conflict between two ethnic minorities that should be brokered, mitigated, massaged. We must reject the notion that leaders who acknowledges anti-Semitism must pay a price for somehow disrespecting their Muslim constituencies. Surely we oppose all forms of bigotry, including anti-Muslim hatred, but exposing anti-Semitism as it is found in our society should not be taken as a denigration of any other religion or group.

We hear much about controversy surrounding the identification of perpetrators and have seen examples of how naming sources of anti-Semitism is considered too provocative. Those who oppose identifying sources and perpetrators think exposing anti-Semitism should be limited by a fear of insulting the communities to which perpetrators of hate violence belong.

For those of us who have watched the problem closely, it is without question that key factors enabling the growth of anti-Semitism have been the fear, reticence, and inability to talk about it in honest terms. Something about defining and talking about anti-Semitism today touches a raw nerve. As with any disease, the denial is insidious and makes it fester and grow. One cannot talk about anti-Semitism in the OSCE region without confronting the role of the Arab World in propagating the kind of anti-Jewish myths that flourished in Europe centuries ago. For so long anti-Semitism was the elephant in the room that no one dared name, but it loomed large. Today we see some more honest understanding of how ancient anti-Semitic canards are being revived and cloaked in theology and religion. Islamist campaigns within the

Muslim World and Europe have moved the anti-Jewish beliefs within Islam in from the fringes, where they historically resided, closer to the center. This demonization of Jews and Judaism emanates from houses of worship and from clerics. It pervades educational systems and government-sponsored media. In the digital age, it is beamed around the world and permeates popular culture well beyond the Middle East.

Our challenge is to find a way to go forward with the message that it is vitally important to identify the disease without letting that be viewed as a political issue or an insult or a slander to any other religion or ethnic group.

The conference ended on a positive note when a "Berlin Declaration" against anti-Semitism was unveiled, pledging to "intensify efforts to combat anti-Semitism in all its manifestations and to promote and strengthen tolerance and non-discrimination." We are finally moving in the right direction.

One of the most telling events of the past year was the controversy surrounding Mel Gibson and his movie *The Passion of the Christ*. My concern about the choices Mel Gibson was making in bringing his version of the death of Jesus to the big screen was that it could fuel anti-Semitism. I did not believe that there would be pogroms in the streets of New York, Los Angeles, or Miami, or that American Jews would be physically attacked. I worried that the charge "Christ killer," used as the rationale for anti-Semitism for 1,960 years, would be hurled again, and that, especially in Europe and the Arab/Muslim World, already rife with anti-Semitism, it could gain credibility.

Outreach to Mr. Gibson proved futile. One of his choices was clearly to avoid dialogue. Rather, he leveled the false charge that we had called him an anti-Semite, something neither I nor the ADL ever did. He also chose to ignore forty years of Church teachings showing that the Romans not the Jews were responsible for Jesus' crucifixion. Rather, he based his film on selective sections of the Gospels and the teaching of an anti-Semitic nun.

The story of the Passion can be told without disparaging the Jewish people. Such an account is mandated by the Catholic Church as a result of the Second Vatican Council, which in 1965 repudiated both the deicide charge and all forms of anti-Semitism in its document, *Nostra Aetate*. Most Protestant churches followed suit, and since 1965 Christians have worked cooperatively with Jews to correct anti-Semitic interpretations within Christian theology. Aside from theological considerations, artists have a moral and social responsibility to avoid promoting material that may foster hatred, bigotry, and anti-Semitism.

Mel Gibson also decided to preselect his audience by reaching out with private screenings to the Evangelical Christian community and like-minded political conservatives. All our requests for a screening were in vain.

However, I did manage to see a late cut on January 21 in Orlando, Florida. The ADL's Director of Interfaith Affairs, Rabbi Gary Bretton-Granatoor, registered to attend a convention of Evangelical pastors at which a screening was part of the program. I registered as well, under my own name, but as an assistant to Rev. Bretton-Granatoor. That is how I saw the film, along with five thousand others, all Evangelical Christians but for a few journalists and a handful of Jews who managed to be registered for the conference.

What I saw confirmed my worst fears. Viewing the film through Jewish eyes, it was clear who Mr. Gibson blamed for Jesus' death. He portrayed the Temple priests and the Jewish crowd as the instigators and the ones with the power to affect the crucifixion, while portraying Pontius Pilate as a sympathetic character who would do anything not to crucify Jesus.

I continued to give Mr. Gibson the benefit of the doubt, as he said the film was still "a work in progress," leading up to its commercial release on Ash Wednesday, February 25, 2004, hoping he would react to our concerns. If he couldn't change the film, I called on him to add a pre-script or a postscript saying something similar to what Cecile B. Demille added in 1928. DeMille decided to revise his film *The King of Kings* after hearing concerns from Jews, Catholics, and others. He added a foreword to his film in which he explained that the Jews, then and now, should not be held responsible for the death of Jesus.

On Wednesday, February 25, my colleagues and I went to the first showing of *The Passion* at a local New York movie theater. My fears were not allayed. I saw an unambiguous portrayal of Jews as responsible for the death of Jesus. At every single opportunity, Mr. Gibson's film reinforces the notion that the Jewish authorities and the Jewish mob are the ones ultimately responsible for the Crucifixion.

The film represents a setback to more than forty years of Jewish-Christian relations. Yet as problematic as it is, the negative consequences can be contained. The last forty years of Church teaching have eroded the base of anti-Semitic thinking among many Christians. We were heartened that since the controversy began, some Christian leaders, including the National Conference of Catholic Bishops and ecumenical and evangelical Protestant leaders, have stood up for the new Church teachings about Jews.

The film became a blockbuster, breaking all records. I am often asked, didn't I contribute to making it a hit by fueling the controversy and public debate? My answer is simple. I—and all Jews—do not have the luxury to sit quietly by in the face of potential anti-Semitism.

But we have received reports of children being called Christ killers in the schoolyards and classrooms. Virulent anti-Semitic hate mail continues to pour into the ADL. Even non-Jewish journalists, columnists, film reviewers, and editorial cartoonists critical of the film (it received mostly negative reviews) have also reported receiving anti-Semitic hate mail. And the film has proven to be fodder for traditional anti-Semitic extremists, making their case for them. In a poll of 1,703 Americans conducted March 17-21, after many people had seen the film, 26 percent said "yes" when asked, "Were the Jews responsible for Christ's death?" That translates into tens of millions of Americans, a legitimate cause for concern.

It has been said that Mr. Gibson's film represents the greatest tool for evangelization that has ever existed. Indeed, more people will see this film than all the Passion Plays from the Middle Ages to the present. It is not just the film in movie theaters that has raised our concern. The DVD copies (no doubt with additional footage and deleted scenes) are being rushed out for an August 31, 2004, availability date. These will be shown in youth gatherings, religious schools and other places without regard to modern scholarship and teachings.

Would I speak out in this same way again? Yes. Would I handle it differently? Perhaps. What is clear to me is that hatred of Jews still resonates with too many people, and as long as I can I will do all I can to shine the light of day on anti-Semitism in hopes to educate others and keep this hatred unacceptable and at bay.

Why This Book—And Why Today

WE LIVE IN TUMULTUOUS TIMES. The continuing war on terror, unrest in the Middle East, and a faltering world economy are capturing headlines everywhere and contributing to a widespread sense of unease and anxiety.

Through it all runs a disturbing current of which many people are only dimly aware. Consider a few of the news stories of recent months that probably did *not* appear in your local newspaper:

Moscow, Russia (May 27, 2002): A young women is injured in an explosion near the Kiev highway while trying to remove a booby-trapped roadside sign reading "Death to the Jews!"

Wittstock, Germany (September 5): A museum honoring the victims of the Nazi death march is firebombed. Outside the museum, a swastika, SS symbols, and an anti-Semitic slogan are painted on a nearby memorial.

Paris, France (January 3, 2003): Rabbi Gabriel Farhi of the Jewish Liberal Movement is stabbed in the stomach by a masked man shouting "Allahu Akhbar." Three days later his car is torched.

Malaga, Spain (early March): The director of the ArtMalaga gallery refuses to stage an exhibit of paintings by Haifi artist Patricia Sasson, saying, "We certainly hold an anti-Semitic attitude to any person related to that country" (that is, Israel).

Sydney, Australia (mid-March): The South Sydney Synagogue is attacked by an unknown arsonist.

Brussels, Belgium (March 19): A Molotov cocktail is thrown at a synagogue by unknown perpetrators, damaging the entrance door.

Paris, France (March 22): Two Jews are stabbed by demonstrators, apparently Arab North African immigrants, who had participated in an antiwar demonstration.

Baku, Azerbaijan (May 3): Camilla Krichevsky is stabbed to death by a Muslim neighbor, who reportedly confesses to police that he believed he would go to heaven if he killed a Jew.

Encino, California (May 8): On Israel Independence Day, a fire is started in the sanctuary of the Valley Beth Shalom synagogue, possibly by a Molotov cocktail thrown into the building.

Allentown, Pennsylvania (July 3): A Molotov cocktail is thrown at Temple Beth El.

There have always been people who hate. Perhaps there always will be. Hate crimes like the ones I've listed may never be completely eradicated. But when I consider this list of incidents occurring within just a few months—and this is just a sampling from among the *hundreds* that took place—I am deeply troubled.

We are not talking about a handful of incidents perpetrated by a few isolated individuals but rather about a little-noticed, under-the-radar pattern of repeated attacks, often violent, occurring in country after country. The pattern is particularly strong in Europe, the historic breeding ground of anti-Semitism and the continent where, just sixty years ago, the greatest crime in history—the Holocaust—was perpetrated against the Jewish people. But the same pattern can be seen in the Middle East, Latin America, parts of Africa, and even in the world's most tolerant nation, the United States.

The reemergence of worldwide anti-Semitism is a phenomenon of the past several years. It began during 1998–1999. Then in 2000, the worrisome trend exploded: major violent attacks on Jews more than doubled. The pattern has continued since then, as illustrated by the litany of violence cited above.

Why is this happening now? That's a question I'll explore in the pages of this book. There are several possible explanations.

In a number of countries, including Belgium, France, Russia, Hungary, Romania, Chile, and Brazil, the extreme right experienced a political resurgence during 1998–1999. Such rightward shifts have coincided traditionally with an increase in anti-Semitism.

The deterioration of the peace process in the Middle East, culminating in the outbreak of the Palestinian *al-Aqsa intifada* ("the second *intifada*") in September 2000, is undoubtedly a major factor.

Finally, the generally dismal economic climate around the world has contributed to a sense of malaise and hopelessness among many of the disaffected. History suggests that economic distress often finds an outlet in hatred. (It's no coincidence that history's most catastrophic outbreak of anti-Semitism occurred in Europe in the 1930s, during a worldwide economic depression.) It's likely that all of these factors have combined to produce today's dangerous trends.

For decades Jewish leaders and others of goodwill have repeated the litany "Never again!" It has been a rallying cry and an expression of our determination that the horror of genocide will never be repeated. Now I find myself forced—to my shock and dismay—to add a question mark to the phrase: "Never again?"

In my lifetime I never expected to witness hatred reemerging so boldly from the darkness. I had hoped and believed that the world had learned something from the horror of six million people—including one and a half million children—being slaughtered solely for the crime of being Jewish. There were reasons to believe that the world was changing, attaining a new level of understanding and tolerance with each passing generation. In particular, Europeans were beginning to seriously grasp and grapple with their responsibility for the past and, more important, for the future.

Now those positive trends are moving in reverse. I am convinced we currently face as great a threat to the safety and security of the Jewish people as the one we faced in the 1930s—if not a greater one.

This may be a shocking claim. Perhaps I sound like an alarmist. I pray that I am wrong. But I speak advisedly, after long and careful study of the historical record, the world's current political and social climate, and the frightening contours of what I call the new anti-Semitism.

Within living memory, we've seen what can happen when a nation or a continent experiences an unrestrained outbreak of anti-Semitism. The Jews of the world—and all people of goodwill who share their desire for a just and free society—learned a series of critical lessons from the tragic history of the twentieth century. Today we understand how important it is to recognize the emergence of new forms of anti-Semitism so that we can warn the world and stave off the worst effects.

That is why I have written this book now: because the early warning signs of anti-Semitism are more troubling today than at any time since World War II. We can't afford to wait for the next crisis and the explosion of hatred and violence it may provoke. Instead, we must act now to prevent that outbreak before it occurs—and perhaps, God willing, to wipe out the threat of anti-Semitism once and for all.

One important note. Many individuals and groups are mentioned in this book. I criticize some of them sharply. But not all deserve the label "anti-Semitic." I hope that in each case the full context will make my opinions unmistakably clear. We owe it to one another to form such opinions thoughtfully and weigh our language carefully. Maintaining an attitude of mutual respect—and above all, of respect for the truth—is also part of the struggle against bigotry and intolerance.

CHAPTER 1

As Storm Clouds Gather:

The Rise of the New Anti-Semitism

IN SOME WAYS TODAY'S REEMERGENCE of anti-Semitism in a time of turmoil is predictable and unsurprising—although it's profoundly disappointing to realize how quickly millions of people seem to have forgotten the lessons of the Holocaust. What's most troubling is the way today's new anti-Semitism combines traditional bigotry and hatred with modern resentments in a way that is unprecedented and particularly virulent. The process is furthest advanced and most visible in the Middle East and on the continent of Europe, but it can be observed also in the United States, in Latin America, and in other countries around the world—even those, like Japan, in which very few Jews can be found.

STRANGE BEDFELLOWS

In today's new mutant strain of anti-Semitism, traditional elements of the extreme right and the extreme left are working together, often in concert with immigrants of Arab descent and terrorist organizations based in the Middle East. It's a strange alliance, since adherents of the far right in Europe and North America generally espouse nativist policies and abhor foreign, especially non-Christian, influences. Yet hatred of the Jews is proving to be a powerful enough force to unite these disparate groups.

The authoritative *Anti-Semitism Worldwide,* prepared annually by a team of scholars at the Stephen Roth Institute for the Study of Contemporary Anti-Semitism and Racism at Tel Aviv University, provides a wealth of details concerning how fringe groups on the right and the left are uniting on behalf of an anti-Semitic agenda. Here are a few examples from the most recent reports concerning the continent of Europe:

- In Germany, Hans-Günther Eisenecker, vice chairman of the radical right-wing NPD, has described his vision of an "antisemitic internationale," which would link Islamist movements, North Korea, Cuba, and extreme nationalists in Europe and America in an alliance against Israel and the United States.

- American white supremacist David Duke recently returned from a two-year tour of Europe, during which he lectured on "the Aryan race's main enemy, world Zionism" in several countries and conferred with Russian leaders ranging from neofascist Vladimir Zhinovsky to Communist Duma member General Albert Makashov.

- In eastern Europe, the right-wing Greater Romania Party and the Hungarian Justice and Life Party have championed the cause of Iraq and the Palestinians, claiming that American foreign policy is controlled by Israel and the Jews.

- Extreme left-wing and right-wing groups worked together to mount anti-Israel demonstrations in autumn 2000 in Rome, Copenhagen, and Vienna.

- Over 250 violent anti-Semitic incidents were perpetrated through Europe in the weeks following the outbreak of the Middle Eastern *intifada,* aimed not at institutions identified with the state of Israel, but simply at Jews. Evidence shows that some of these attacks were conducted by left-wing groups, others by right-wing groups, and that the two sets of extremists appeared to inspire each other in anti-Jewish attacks.

Similarly, in the United States, white supremacists and other ultra-right-wing groups have taken up the Palestinian cause in an effort to build a new coalition of hatred to battle the Jews:

- Matt Hale, leader of the white supremacist World Church of the Creator, has called on his supporters to ally themselves with the Palestinians against the common Jewish enemy. In an April 10, 2002, press release, he called suicide bombing "an obviously effective technique that courageous Palestinians in their determination to expel the Jewish invader of their lands have decided to employ."

- In a similar vein, David Irving, a historian whose stock in trade has been the traditional ultra-right-wing denial of the Holocaust, has written on his Web site of the "suicidal heroism" of the "Arab world" and denounced Israeli prime minister Ariel Sharon as a "terrorist."

- The notorious David Duke, the former Ku Klux Klan leader who has tried to import his brand of bigotry into the

Republican party, has been giving pro-Palestinian speeches attacking "Jewish Supremacist Chutzpah" and ridiculing "the so-called holocaust."

Meanwhile, left-wing groups in the United States, especially on college campuses, have taken up the anti-Israel cause and pushed it over the line into outright anti-Semitism. The frightening incidents have swiftly mounted in recent months. Here are some examples:

- At protest marches against the 2003 war in Iraq organized by Action Now to Stop War and End Racism (ANSWER), anti-Jewish and pro-terrorist slogans and speeches abounded, from "End the Holocaust" (with a picture of Israeli prime minister Ariel Sharon) to "First Jesus, Now Arafat—Stop the Killers." At a San Francisco rally organized by ANSWER on February 15, 2003, Rabbi Michael Lerner, editor of the leftist magazine *Tikkun,* was forbidden to deliver an antiwar speech because of his pro-Israel position.

- At an anti-Israel rally at the University of Denver, speakers compared Zionism to Nazism, and a member of the Colorado Campaign for Mideast Peace called a Jewish student a kike. On the same day, at another rally at San Francisco State University, posters were displayed bearing pictures of soup cans labeled "Made in Israel. Contents: Palestinian Children Meat."

- An associate professor of history at Kent State University in Ohio used his column in the campus newspaper, the *Kent Stater,* to praise a female Palestinian suicide bomber.

- Muslim student groups at the University of California's Berkeley and San Diego campuses posted flyers featuring bogus anti-Semitic quotations from the Talmud, including statements such as "A Gentile girl who is three years old can be violated" and "When the Messiah comes, every Jew will have 2800 slaves."

· At Illinois State University, a Jewish student who was asked to sign a petition in support of Palestinian rights asked whether the petition addressed the issue of suicide bombings. In response, a Palestinian graduate student said the petition talked about how to blow the Jewish student's head off.

· Pro-Palestinian demonstrators at San Francisco State University screamed slogans including "Go back to Russia" and "Hitler did not finish the job" at Jewish students expressing support for Israel.

In such cases, the traditionally right-wing language and imagery of anti-Semitic hatred is being used to subvert the ostensibly progressive left-wing anti-Israel cause of peace and equality in the Middle East. There are many reasons why this new unholy alliance between far right and far left is particularly troubling.

First, the multiple tentacles of the new anti-Semitism provide access to many social and political groups that otherwise would be difficult or unlikely targets for the spread of bigotry. For example, left-wing anti-Israel activists are likely to obtain a sympathetic hearing among college students, socially conscious church organizations, and environmentalists, while right-wing activists are able to connect with veterans' organizations, businesspeople, and conservative religious groups. Thus the new anti-Semitism is capable of reaching people who would be unmoved by such traditional anti-Semitic themes as xenophobia and religious prejudice.

The new anti-Semitism also makes bigotry appear more mainstream, more sophisticated, and therefore more acceptable than in the past. A youthful liberal who would find traditional anti-Semitism old-fashioned and easy to discredit is much more likely to find the new anti-Semitism appealing, cloaked as it is in the more fashionable rhetoric of anti-imperialism, anti-racism, and anti-Americanism.

Finally, the new anti-Semitism has multiple sources of financial and logistical support. These include not only sympathetic supporters in Europe and America but also well-heeled individuals, businesses, and governments in the Arab and Muslim worlds. The complex interconnections among the many groups that foster the new anti-Semitism make it extremely difficult for opponents to trace, expose, and block the sources of this support.

Thus the new anti-Semitism isn't just a recycling of the same old hatred, being peddled by tired and long-discredited spokesmen for bigotry. Instead, it's a new form of poison that blends several streams of intolerance into a particularly deadly cocktail.

STORM WARNINGS

Monitoring worldwide developments in this deepening crisis has become an urgent mission. In October 2002 the Anti-Defamation League convened an International Conference on Global Anti-Semitism attended by many leading diplomats as well as Jewish community representatives from around the world. The reports presented at this conference confirmed many of our worst fears.

We heard from Roger Cukierman, president of the Conseil Representatif des Institutions Juives de France (CRIF), who described the "general feeling of vulnerability and anxiety" that French Jews had lived with since the start of the *intifada*. In France, as elsewhere in Europe, anti-Semites have carefully blurred the line between criticism of Israel and hatred of Jews. "French public opinion, in general, has an extremely negative view of the state of Israel, and this attitude carries over to French Jews," Cukierman explained. Complicating the situation in France is the growing political power of Muslims, who number over four million and are increasingly influenced by religious fundamentalism.

The atmosphere in Britain is almost as bad, as Jo Wagerman, president of the Board of Deputies of British Jews, explained to us at the same conference. Until recently, anti-Semitism in Britain was restricted to minority and insignificant splintered groups on the far right. But now the United Kingdom is facing an explosion of anti-Semitism, especially through its Muslim population. Arab newspapers, TV and radio stations, books, films, popular songs, children's programs, political speeches, and Muslim clerics put out a stream of hatred. Their relentless message is that Jews, who are the origin of all evil and corruption, plan world domination and the eradication of Islam and Christianity. The aim of such propaganda is not simply to delegitimize Israel as a Jewish state but to dehumanize Judaism and the Jewish people.

There is widespread targeting, intimidation, and harassment of Jewish students on British campuses. There is also an ongoing campaign in support of a boycott of Israeli businesses, which is taking on an increasingly anti-Semitic nature. Meanwhile, the media's obsession with blaming Israel for the current impasse in the Middle East has created an environment in which anti-Semitism is increasingly accepted. "Israel is the scapegoat. Jews are blamed as the agents of global imperialism and Westernism," Wagerman said.

The new anti-Semitism is also obtaining a foothold in the Americas. Alfredo Neuburger, director of the Political Affairs Division of DAIA, a Jewish umbrella organization in Argentina, reported that the Jewish community in Buenos Aires is still reeling from the effects of two deadly terrorist bombings in the 1990s, one targeting the Israeli embassy (1992), the other a Jewish community center (1994). Recently these attacks have been traced to Hezbollah, the Lebanese-based terrorist organization, working with the help of the Iranian government.

In retrospect, these bombings can be seen as precursors of the September 11 attacks in the United States—attacks by Islamic terrorists on civilian targets distant from the Middle East, selected because they symbolize aspects of the Western world that are anathema to the extremists (Israel, Jews, and American capitalism). Unfortunately, the world failed to heed these warning signs.

As suggested by reports such as those of Cukierman, Wagerman, and Neuburger—seconded by other observers from countries all over the world—today's time of crisis finds fertile ground for the resurgence of ancient anti-Semitic prejudices laid down over centuries by hate-mongers, propagandists, and leaders of church and state. The traditional anti-Semitic beliefs that had seemingly been on the wane prior to the current *intifada* and the attacks of September 11 have now achieved a new life and are blending with the anti-Israeli, pro-Muslim strains that help make up the new anti-Semitism.

THOUGHTS THAT CAN KILL

At the Anti-Defamation League, we make it our business to monitor the strength, persistence, and spread of anti-Semitic attitudes. Since 1964 ADL has conducted a number of public opinion surveys in the United States to measure levels of anti-Semitism. An index of eleven questions was developed by researchers at the University of California to be used in these public opinion surveys. It serves as an analytical tool for identifying respondents with a propensity to be prejudiced against Jews.

Before answering the index questions, respondents are read the following statement: "I am now going to read out a series of statements. Some of them you will agree with and some of them you will not. Please say which ones you think are probably true and which ones you think are probably false." Respondents who agree with six or more of the eleven statements are classified as "most anti-Semitic."

The following are the eleven statements that constitute the anti-Semitism index:

1. Jews don't care what happens to anyone but their own kind.

2. Jews are more willing than others to use shady practices to get what they want.

3. Jews are more loyal to Israel than to this country.

4. Jews have too much power in the business world.

5. Jews have lots of irritating faults.

6. Jews stick together more than other Americans (or citizens of other countries).

7. Jews always like to be at the head of things.

8. Jews have too much power in international financial markets.

9. Jews have too much power in our country today.

10. Jewish business people are so shrewd that others do not have a fair chance to compete.

11. Jews are just as honest as other business people. [Considered prejudiced if answered "probably false."]

In the latest survey, 1,000 Americans age eighteen and older were interviewed between April 26 and May 6, 2002. Respondents were selected using a random probability sampling procedure. In addition, 300 African American and 300 Hispanic American respondents were selected to increase the reliability of the results obtained within these important subgroups. In addition, for the first time, 800 college students and 500 college faculty were interviewed to measure anti-Semitic tendencies on campuses.

The results were troubling, to say the least. The key finding in the 2002 survey results is an increase in anti-Semitic attitudes among Americans, reversing a decade-old trend. Fully 17 percent of those surveyed hold views about Jews that are unquestionably anti-Semitic, as compared to just 12 percent in the previous survey (1998). This translates into 35 million adult Americans with anti-Semitic views—an enormous base of potential support for bigoted political views and policies.

In 2002 the anti-Semitism index was employed for the first time in Europe. We commissioned a research firm, First International Resources, to survey anti-Semitic attitudes in ten Western European countries: Belgium, Denmark, France, Germany, and the United Kingdom (in June 2002) and the Netherlands, Austria, Italy, Spain, and Switzerland (in September). A total of 5,000 telephone interviews were conducted, 500 in each of the ten countries.

As you might imagine, the results vary by country. For example, the percentage considered "most anti-Semitic" is 34 percent in Spain and just 7 percent among the Dutch. Overall, 21 percent—one out of five European respondents—are characterized as "most anti-Semitic." I consider this figure alarmingly high, especially considering the degree of self-censorship most people will practice when being surveyed on a sensitive topic like anti-Semitism. If a fifth of Europeans are willing to openly admit anti-Semitic beliefs, how many more are secretly sympathetic to such beliefs?

When we focus on individual index items, even more troubling results emerge from the most recent surveys of Europe and the United States:

- Forty percent of Europeans—and nearly a quarter of Americans (24 percent)—believe that Jews have too much power in the business world and in international finance markets.

- Nearly two-thirds of Europeans (63 percent) and 50 percent of Americans believe that Jews "stick together" more than other people.

- Thirty-five percent of Americans and 29 percent of Europeans believe that Jews "always like to be at the head of things."

- Forty percent of Europeans and 20 percent of Americans believe that Jews have "too much power" in financial markets.

- Twenty-nine percent of Europeans and 16 percent of Americans believe that Jews "don't care what happens to anyone but their own kind."

The persistence into the twenty-first century of these classic anti-Semitic beliefs is very disturbing. But perhaps most worrisome of all are the survey results that point to another aspect of the new anti-Semitism: the doubt it casts on the loyalty of Jewish citizens, suggesting that in time of crisis Jews could be singled out for legal or political sanctions as potential traitors.

In response to question 3 in the anti-Semitism index, a majority of Europeans—56 percent—responded that it is "probably true" that Jews are more loyal to Israel than to their own country. In the United States, one-third of respondents (33 percent) felt the same.

In response to other questions, 53 percent of the Europeans surveyed said they believe that the recent outbreak of violence against Jews in Europe is a result of "anti-Israel sentiment," while just 17 percent said they believe it is a result of anti-Jewish feelings. (Nine percent said they feel that both are contributing factors.)

Why is this finding disturbing? Because the belief that violence directed against Jews is a kind of political statement about international conflict tends to legitimize it—as if attacks on synagogues and schools are innocent acts of protest, comparable to demonstrations against apartheid or the war in Vietnam.

ANTI-ZIONISM AND ANTI-SEMITISM

In every public forum, I'm always careful to say that criticism of the state of Israel is not necessarily anti-Semitic. There are legitimate disagreements over the path to peace in the Middle East. Israelis themselves are engaged in an ongoing debate over exactly how to resolve the tension between the aspiration of Palestinians for their own homeland and the essential security needs of Israel.

But a prominent feature of the new anti-Semitism is its use of anti-Israeli politics, sometimes described as anti-Zionism, as a mask for anti-Semitism.

First, a couple of definitions. As I use the term, *Zionism* simply refers to support for the existence of a Jewish state—specifically, the state of Israel. Zionists believe that the desire of Jews to have a homeland of their own in the Middle East is historically based and legitimate. Zionists also believe that Israel, like other nations, has the right to exist in peace, within secure borders, and to use force to defend itself from attack when necessary. Zionism, in effect, is Jewish nationalism, comparable to the nationalism espoused by most other ethnic groups around the world, including those who have their own states (such as the Russians and the Japanese) and those who do not (such as the Basques in Spain and the French-speaking people of Quebec).

Anti-Zionists disagree. They contend that the desire for a Jewish homeland is illegitimate; that Jewish claims to the territory of Israel have no basis in history, tradition, or law; and that the state of Israel should either abandon its legal and political status as a Jewish nation or cease to exist altogether.

Americans are accustomed to thinking that there are two legitimate sides to any dispute. We tend to assume that most conflicts involve opposing parties who take extreme positions and that truth and justice are to be found somewhere in the middle. This may be true in many cases. But the logic of split-the-difference doesn't apply to the conflict between Zionists and anti-Zionists. The harsh but undeniable truth is this: what some like to call anti-Zionism is, in reality, anti-Semitism—always, everywhere, and for all time. Therefore, anti-Zionism is not a politically legitimate point of view but rather an expression of bigotry and hatred.

This is a strong claim, but I believe it's a just one. Consider, for example, the United Nation's notorious anti-Zionism resolution. Passed in 1975 by the UN General Assembly by a vote of 89 to 67, it declared that "Zionism is a form of racism."

Why is this statement anti-Semitic? Think about what the resolution states and what it implies. If Zionism is racism, then what is permissible, what is laudable, what is universally accepted for all peoples in the world—self-expression, self-determination, independence, sovereignty—is not permitted to Jews. The anti-Zionist doesn't condemn Irish nationalism or Basque nationalism or French nationalism or Palestinian nationalism. Only Jewish nationalism is racist. That is a double standard, which is inherently unfair. To uphold such a double standard solely for the purpose of condemning Jews can only be described as anti-Semitism.

If you question my insistence that anti-Zionism is anti-Semitic, don't take my word for it. Consider the opinion of a person who knew a little something about racism, Dr. Martin Luther King Jr. In a 1968 speech at Harvard University he addressed this very question. Here is his conclusion:

You declare that you do not hate the Jews, you are merely anti-Zionist. And I say, let the truth ring forth from the high mountain tops, let it echo through the valleys of God's green earth: when people criticize Zionism, they mean Jews. This is God's own truth. . . . Zionism is nothing less than the dream and ideal of the Jewish people returning to live in their own land. . . . And what is anti-Zionism? It is the denial to the Jewish people of a fundamental right that we justly claim for the people of Africa and freely accord all other nations of the globe. It is discrimination against the Jews, my friend, because they are Jews. In short, it is anti-Semitism.

Dr. King was right in 1968. He's right today.

Thankfully, the "Zionism is racism" resolution was revoked in 1991 by a UN vote of 87 to 25—although this vote of course indicates that, as of 1991, at least twenty-five states were still willing to openly maintain the position that Zionism is a form of racism, thereby seeking to delegitimize Israel and threaten the Jewish right of self-determination.

This same strain of anti-Semitism, thinly veiled as anti-Zionism, can be seen also in the distressing story of the UN-sponsored World Conference Against Racism in Durban, South Africa, in August 2001. The lessons of Durban help illustrate why what is happening throughout the world today is so much more dangerous, so much more sinister, than the forms of anti-Semitism we have had to combat in recent decades.

In the eyes of many of those planning the conference, Durban was to be a magnificent expression of the world community's conscience and care of the future. At the turn of the new millennium, the nations of the world decided that, since the world had paid such a heavy price for racism, let the community of nations come together to set standards for dealing with racism. It was a beautiful concept. And the choice of Durban, South Africa, as the site of the conference was deeply symbolic, since South Africa is where racism in one of its ugliest forms—apartheid—had recently been vanquished. There was a real need for the world to set standards of behavior, and this conference was supposedly intended to address that need.

Yet the attendees at the conference never really got to talk about racism. They never got to set the standards or deal with the issues that so many of those who came from all over the world had hoped to address. One subject dominated the discussion—the Jewish people and Jewish "racism."

We at ADL, who had hoped to contribute to the conference, watched the proceedings with dismay. We saw participants in the Non-Governmental Organization (NGO) Forum and Youth Summit bombarded with statements and demonstrations of hate aimed at Jews, Israel, Zionism, and the United States. With support from conference organizers and panel moderators, NGO speakers ignored their own agendas to digress into Israel-bashing. Jewish students distributing flowers to the delegates while singing the idealistic John Lennon tune, "All we are saying is give peace a chance," were drowned out by a large crowd holding Palestinian flags and banners, denouncing "Israeli apartheid" and hailing the *intifada*. Even a modest attempt by the Jewish organizations from around the world to present their concerns in a press conference in Durban was interrupted by protesters shouting anti-Israel epithets.

Perhaps we shouldn't have been surprised. The moment the community of nations allowed a planning meeting to be held in Tehran, Iran, we should have known the direction Durban would take. What was frightening was that aside from the United States and Israel (both of which walked out of the conference), and a few other countries that issued mild statements of protest, the world permitted the subversion of the conference in support of the cause of delegitimizing the Jewish people. Many otherwise good people from around the world apparently found it impossible to raise their voices, to vote against this travesty, or to walk out of the conference.

Events like Durban make it clear that anti-Zionism is nothing more than the newest mask worn by hatred—a facade of legitimacy used in an attempt to make anti-Semitism welcome in mainstream cultural and political circles, in Europe, in America, and around the world. Israel is frequently judged more harshly, more unforgivingly, than any other nation. When tens of thousands of civilians were killed in Chechnya, when half a million people were slaughtered in Rwanda, when two million Christians and animists were killed in the Sudan, the protests around the world were muted, at best. But every casualty in the Arab-Israeli conflict is trumpeted as an atrocity unparalleled in history.

No wonder I say that most of the current attacks on Israel and Zionism are not, at bottom, about the policies and conduct of a particular nation-state. They are about Jews.

Again, let me stress that criticism of Israel or its government is not in itself illegitimate. But those who offer such criticism should be held to a basic standard of fairness. When other countries and people pursue policies that are similar to (or far worse than) those of Israel, do the critics condemn them? If so, do they condemn them with the same fervor as they condemn Israel? If not, it's hard to deny that anti-Semitism explains the discrepancy.

ISLAM RISING

One more trend must be mentioned as playing an important role in the resurgence of worldwide anti-Semitism: the rapid increase in the Muslim populations in traditionally non-Muslim regions around the world.

Don't misunderstand: I don't mean to imply that all Muslims are anti-Semitic, let alone that all Muslims are apt to commit violence or other crimes against Jews. I abhor attempts to paint any religious or ethnic group with a single brush, and that applies to adherents of the Islamic faith just as it applies to Jews, Christians, or Blacks. I've spoken out vigorously against such stereotyping, and I'll always do so.

But it's a fact that anti-Semitism is rampant in the world of Islam. (I devote an entire chapter to this topic.) And with intolerant, fundamentalist Islamist extremists occupying center stage and dominating public discourse in much of the Muslim world, those Muslims who are predisposed to hate Jews and even to commit acts of violence against them receive plenty of acceptance and even encouragement from their peers.

Now, in a historic shift, the attitudes of the Islamic world are beginning to have an important presence in traditionally non-Muslim regions of the world.

For more than three hundred years (since the historic defeat of a Turkish army by Polish troops near Vienna in the Battle of Chocim), Europe has been a Christian enclave. With the exception of Turkey and the Balkans, few European countries have had significant Islamic populations. Today that is changing. Since the 1950s Muslims have flocked to Europe as "guest-workers," taking advantage of low-level economic opportunities many native Europeans scorn. And with native European populations stagnant or declining, creating labor shortages, the influx has accelerated in recent years.

Now over 15 million Muslims live in the countries of the European Union, and Islam is the fastest-growing faith in Europe. If current trends continue, by 2020 Muslims will represent 10 percent of the population of Europe. At the same time, Jewish populations in Europe have declined, as Jews continue to emigrate to Israel and the United States. For example, between 1961 and 2001 the Muslim population of Britain rose from 82,000 to 1 million, even as the Jewish population declined from 400,000 to 250,000.

In the United States, the same trend is apparent, though it is less advanced. There are already several million American Muslims, of whom about three-quarters are immigrants to these shores. The numbers of American Muslims are increasing rapidly, due to both immigration and conversion.

Under the circumstances, it's not surprising that politicians, especially in Europe, are coming to regard Islamic voters as a burgeoning power bloc to be reckoned with. Of course, European political leaders are not being forced to tolerate or adopt the anti-Semitic attitudes of many Muslims. But unfortunately, as the political pressures from the growing Muslim population mount, few European politicians have had the courage to denounce and actively discourage the spread of anti-Semitism. Most find it easier to accommodate, excuse, or ignore anti-Semitism, even when it takes the form of violent crime in their own countries.

COWARDICE AND EXPEDIENCY

Many European governments have been less than forthcoming in their statements regarding the upsurge in anti-Semitic violence. The French government, for example, has yet to release official statistics on such incidents in 2002. When politicians do address the problem, their statements are often dismissive or evasive. For instance, when a French government spokesman belatedly acknowledged (in June 2002) that "a series of inexcusable assaults—physical, material, and symbolic—has been committed in France against Jews over the past twenty months," he went on to suggest that this was simply a spillover of the Middle East conflict into Europe and blamed most of the incidents on "poorly integrated youths of Muslim origin who would like to bring the Mideast conflict to France." The role of traditional anti-Semitic attitudes and the active involvement of extreme European nationalist groups in anti-Semitic violence were ignored.

Thus, even as synagogues were being burned and Jewish children were being attacked, President Jacques Chirac and Prime Minister Lionel Jospin leaned over backward to attribute these attacks to controversy over the Palestinian-Israeli conflict. Chirac even repeatedly insisted, "There is no anti-Semitism in France." Perhaps this was understandable; after all, Chirac and Jospin faced elections in which several million Arab votes were at stake. It was understandable—but it was craven and dishonest.

As a result of the cowardice of many mainstream European politicians, openings are created for politicians on the fringe to deliberately exacerbate and exploit anti-Semitic attitudes. On April 21, 2002, Jean-Marie Le Pen, the anti-Semitic leader of the far-right National Front, gained second place in the first round of the French presidential election with 17 percent of the total vote (as compared with about 20 percent carried by incumbent Chirac). He thereby qualified for a second-round runoff against Chirac. Thankfully, in the second round Chirac routed Le Pen by a margin of 82 to 18, although it's far from comforting to see a notorious anti-Semite installed as a permanent fixture of French politics, with steady support from 15 to 20 percent of the populace. (Recently, Chirac and French minister of education Luc Ferry have begun to speak out against anti-Semitism and have launched a program of education in tolerance for French school children.)

We see similar movements gaining a foothold in other European countries. In Greece in 2000, ultra-right-wing members of parliament sympathetic to the Greek Orthodox Church blamed Jews for the socialist government's decision to delete information about religious affiliation from state-issued identity cards. (In fact, the change came at the request of the European Community.) Following a huge demonstration orchestrated by the church and extreme right-wing circles, the small Jewish community in Greece suffered intimidation and vandalism.

In Romania the anti-Semitic Greater Romania Party became the second largest party in the parliament, with 21 percent of the vote, following the general election in November 2000. Anti-Semitic violence followed close on the heels of the political results. In December two visitors to the Jewish Historical Museum in Bucharest demanded to see "Auschwitz soap" with their own eyes, then choked and seriously injured a security guard and vandalized the premises.

Anti-Semitism has become a major political weapon of the

nationalist and the communist opposition in Eastern Europe and the former Soviet Union. The new Russian president, Vladimir Putin, has restricted the activities of the extreme right and has tried to crack down on anti-Semitic violence. He even led a ceremony in tribute to Tatyana Sapunova, a courageous young woman who was injured by a bomb while trying to remove an anti-Jewish sign near a road outside Moscow. However, hundreds of anti-Semitic publications can be openly purchased in Russia, and Jewish leaders are concerned about the future direction of Putin's authoritarian regime.

In some European quarters, anti-Semitism has become part of the political mainstream. Shockingly, the seat of the International Court of Justice at the Hague has repeatedly sought to indict the prime minister of the state of Israel for crimes against humanity. The effort represents a cynical and horrifying abuse of the hallowed Geneva Convention, which codified rules of war and behavior and established standards for human behavior even in times of warfare.

The keeper of the Geneva Convention is Switzerland, which must be petitioned for the convention to be called into session. From 1949 to 1999, nobody ever called that convention into session. Of course, from 1949 to 1999 there were certainly atrocities committed against civilians throughout the world. Nonetheless, nobody thought it would be possible or necessary to call the convention into session until 1999, when the Fourth Geneva Convention convened—over what? The "atrocity" of Har HaHoma, a neighborhood in Jerusalem where Israel decided to build housing. It didn't matter that one-third of the housing was set aside for Palestinians. The Fourth Geneva Convention was gathered for the first time in history to denounce the crimes of Israel. And lest the point be misunderstood, the Fourth Geneva Convention met again in 2002. Again, the topic was Israel and its supposed war crimes in the Palestinian conflict.

Such behavior on the part of European statesmen is anti-Semitism, pure and simple. That would not be the case if Belgium, the cradle of international justice, set out to indict other leaders of the world who had committed real atrocities. (A list of candidates for such indictment would not be difficult to compile.) If Belgium were interested in setting a genuine standard of justice for the world, I would be among the first to applaud. But no one else is being charged with war crimes, only the prime minister of Israel.

The same pattern is played out in other incidents scattered across the political landscape of Europe.

In August 2001 Israel appointed a new ambassador to Denmark, Carmi Gilon, who formerly served as head of Shin Beth, the Israeli security service. Danish politicians raised an outcry over Gilon's background, feeling it associated him with "inhumane" treatment of Palestinian prisoners. (Of course, the Danish government has had no difficulty in dealing with Russia's Vladimir Putin, former head of his nation's KGB, or for that matter with America's George H. W. Bush, former head of the CIA.) That is anti-Semitism.

Daniel Bernard, the French ambassador to Great Britain, was quoted as referring to "this shitty little country, Israel," in a December 2001 conversation with the wife of media baron Conrad Black. The context: "a conversation about the Middle East crisis." Bernard's meaning was clear enough: if only "this shitty little country" of Israel didn't exist, there would be no problems in the Middle East. That, too, is anti-Semitism.

Again, there are crucial distinctions that must be drawn. The ambassador is certainly free to criticize Israeli policies—regarding settlements in the occupied territories, for example. There is debate within Israel about what policies ought to be followed. But the ambassador is not criticizing Israeli policies. Neither is he interested in applying the same standard of fairness to countries around the world. It wouldn't be anti-Semitism if the ambassador were equally vocal about the Chinese occupation of Tibet or the

Indian occupation of Kashmir. But it is only Israel and the Jewish people who are singled out for attack. The double standard is what marks the incident as anti-Semitism.

Not every government in Europe deserves equal blame for abetting the rise of the new anti-Semitism. A few are taking principled stands. Spain, for example, is considering working with ADL to develop a program to combat anti-Semitism among its people. (As you saw, our survey of European attitudes found that Spain is one of the hotbeds of anti-Semitic attitudes in Europe.) The Germans and Italians have at least issued strong public statements condemning anti-Semitism. As we've seen, Putin and the Russians have taken some of the right kinds of positions. But too many have followed the lead of the French, trying to sweep the problem under the rug while winking at or even actively condoning anti-Semitic attitudes and actions.

DANGER SIGNS

The rise of the new anti-Semitism is occurring in a context of historic conditions and trends that deepens the danger to worldwide Jewry.

One of the reasons for my concern is that today 40 percent of the Jewish people are centered in a single geographic location, the state of Israel. It's a small nation surrounded by a large collection of hostile neighbors, some with access to weapons of mass destruction, determined to wipe it from the face of the earth. If the Jews of Europe were physically vulnerable in the era of the Nazis, the Jews of the Middle East may be even more vulnerable today.

This worry of mine may seem ironic. After all, wasn't the state of Israel founded in 1948 precisely to serve as a safe haven, a refuge for the world's most beleaguered people? That's true, and Israel has served that function well. Today it remains a heavily armed country blessed with one of the world's best-led and most highly skilled military forces as well as the support of the United States, the most powerful nation on Earth. Until today, that combination of resources has sufficed to protect the country and its people.

But I wonder how much longer the bulwark that is Israel can remain intact. I see dangers looming on several fronts.

First, there is the emerging demographic challenge arising within Israel. A growing fraction of its citizens are Palestinian Arabs, primarily Muslims, whose allegiance to the Jewish state is dubious at best. Because Israel is a democracy—the only democracy in the region—the Palestinians have the vote, a presence in parliament, and an increasing degree of political power. If the West Bank and Gaza remain part of Israel, eventually Israel's non-Jewish population will become an absolute majority.

It's impossible to predict the upheavals this will bring, but it seems clear that, broadly speaking, there are only two possibilities: either Israel will cease to be a democracy, or it will cease to be a Jewish state. After all, if Israel adheres to its democratic principles, when non-Jewish citizens become a majority of the voting population, they will have the power to change the laws and policies of the country so as to completely secularize Israel, removing any vestiges of Jewish heritage from its official practices. The only way to prevent a non-Jewish majority from taking such steps (if they wish to do so) would be to limit their political power—which means a diminution of democratic practice.

I find both alternatives dismaying. Whether Israel loses its identity as a haven for world Jewry or abandons its democratic principles, the shining hope that Israel has long represented for the Jewish people will be radically changed—and not for the better.

I also worry about the possibility that Israel's alliance with the United States could break down, perhaps leaving Israel to defend itself alone. Only a decade or two ago this idea seemed unthinkable. In today's world I'm afraid it's all too possible.

Consider the pressures on the American people and government during these first years of the twenty-first century:

The United States and world economies, already reeling from recession, are heavily dependent on Arab oil. This dependency creates a natural wedge between American and Israeli interests—one that past American administrations have managed to overcome but against which there can be no future guarantees.

As I write these words, in the immediate aftermath of the American victory over the Iraqi regime of Saddam Hussein, the United States is deeply absorbed in the global war on terror and in an ongoing conflict against despotic regimes in the Middle East. These are battles in which Israel has long been on the front lines although frequently out of sight. For the moment, American interests in both conflicts appear to run parallel to those of Israel. But they may not always do so. To offer just one scenario: suppose the United States becomes embroiled in another war in the Middle East. If Israel is forced into that war by an Arab attack, will the war metamorphose into a war between Islam and the West over the survival of Israel?

As the world's sole remaining superpower, the United States is pulled in many directions by interests around the globe, ranging from Southeast Asia to Latin America, from China and India to Russia and southern Africa. American voters are notoriously fickle in their interests—and so are many American politicians. In an era when crises may erupt anywhere at any time, how secure can Israel be in relying on the steadfast support of its American ally?

In the world of the twenty-first century, shaken by the new fears and turmoil of the war on terror, no alliance can be assumed to be permanent, not even an alliance as profound and historically based as that between the United States and Israel. I'm worried about the long-term security of a country that must rely heavily on the support of any outside power, even when that power is my own American homeland.

This, then, is the first reason for my anxiety for the Jews of the world—my concern about the increased vulnerability of Israel.

A second reason is the emergence over the past decade of a global communications system that facilitates the spread of anti-Semitic hatred. Epidemiologists know that one of the prerequisites for a terrible pandemic is the existence of physical and social conditions—crowded tenement housing, for example—that is suitable for the rapid transmission of a virus or bacterium from one person to another. In today's world the electronic communications technology provides the conditions in which the virus of anti-Semitism can be spread instantaneously across a city or around the world.

By contrast, in the 1930s the venom and hatred of Nazi anti-Semitism was mainly limited to Germany and Austria, with a lesser impact in Italy and Spain. True, the agents of Nazism dreamed of spreading anti-Semitism throughout the world, but their power to do so was limited. Although Fascist movements infected other European nations, they never attained significant political power elsewhere in Europe, to say nothing of other regions of the world.

Today the great technological revolution of satellite broadcasting and the Internet has changed the equation—for good *and* ill. On the one hand, global communication provides knowledge, information, education, and enlightenment. It helped pave the way for the peaceful dissolution of the Soviet empire and today may be opening doors to democracy in China, Iran, and elsewhere. On the other hand, the same technology provides a cheap and powerful electronic superhighway for hate. A sermon delivered in Cairo by a virulently anti-Semitic cleric can travel across the globe within seconds via the Internet, e-mail, and the Arab satellite television networks, such as al-Jazeera. So can terrorist instructions for constructing a homemade bomb or the latest hate-filled cartoons and articles from the publications of the extreme right.

Thus the phenomenon of globalization, which we Americans

usually think of as benign, also facilitates the incitement to hate that makes the message of anti-Semitism more potent than ever. It is now out there, everywhere. You can download it—it may even come into your home uninvited.

In America, hate speech, like other forms of communication, is legally protected by our Constitution and our tradition of tolerance. I hope that never changes. But today's technology has given anti-Semitism and other forms of bigotry a strength and a power of seduction unprecedented in history. This places an enormous burden on people of goodwill to fight back—not through violence or even censorship, but by responding unflinchingly and unceasingly to expose the emptiness at the core of the hate-mongers' ideology.

TIMES OF CRISIS AND THE BIG LIE

The electronic network of hatred connecting anti-Semites in every country has been developed over the past decade to the point where it was waiting, as the new millennium began, for the proper confluence of events to kick it into high gear. Now, in the three years since 2000, we have seen those events come to pass. The current array of global crises is an additional reason my latent concern has now grown into real alarm.

The first of these events is the heightened level of unrest in the Middle East. Ever since the new *intifada* began in September 2000, the incidence of anti-Semitic rhetoric and physical violence in countries around the world has escalated enormously, fueled by anti-Israeli feeling.

Those who consider themselves liberal or left-wing around the world, including in Europe and the United States, often lean toward the Palestinian side in this conflict. They do so because of their traditional antipathy to imperialism as practiced by the great Western powers, including the United States,

Britain, and France. This leads them to favor what they perceive as the Third World side in any international dispute. In this case, that means the Palestinians, who are indeed a poor, dispossessed people deserving of sympathy (and of far better treatment than they've ever received at the hands of the Egyptians, Jordanians, and other Arab nationalities that publicly espouse their cause).

Most of these liberal sympathizers with the Palestinian cause denounce the terrorism practiced by some Islamic extremists (although a few excuse it). But the liberals seem to reserve their harshest criticism for Israeli policies. Many overlook or rationalize the use of violence by Palestinians and their supporters against innocent Israeli civilians while denouncing in extreme terms any retaliatory or defensive moves by the Israeli government and military. Because it is so one-sided, this left-wing criticism of Israel often crosses the line into blatant anti-Semitism.

And when verbal attacks on Israel and the Jewish people win mainstream acceptance, it's a small step to excusing or ignoring acts of physical intimidation or even violence directed against Jews, not just in Israel but around the world. This is the social and political backdrop for many of the hate crimes I listed earlier this book.

So the advent of the new *intifada* in 2000 helped create an atmosphere in which anti-Semitism seemed somehow acceptable to many otherwise well-intentioned people. Into this already-worsening climate erupted the horrors of September 11.

For every lover of peace and freedom, the attacks of that day were an unbearable tragedy. But for the Jews of the world, September 11 was especially frightening.

For many years I said that my greatest nightmare was that one day I would wake up to hear that something terrible had happened—and that the Jewish people were to blame. I didn't have any particular disaster in mind; it could have been a financial depression, a devastating epidemic, even an awful natural catastrophe. Such things happen. But what I feared was that the enemies of the Jews would somehow, some way, manage to concoct a story that pinned the blame on us—and that people would listen, and believe,

Was this purely a paranoid fantasy? I wish I could think so. But history teaches otherwise.

When fourteenth-century Europe was being decimated by the Black Plague, fear and hatred fueled by a thousand years of anti-Semitic theology led ignorant people to seize on the Jews as scapegoats. "They have poisoned the wells," was the cry. "Kill them!" And thousands were duly slaughtered.

Such a thing could never happen today, you might think. But fast-forward by half a millennium. In 1997 Malaysia suffered a severe economic crisis; its currency was devalued, and unemployment shot up. The government had no answers. So Malaysia's prime minister, Mahathir Mohamad, sought a scapegoat instead. He blamed the Jews, who supposedly control the world financial system: "We are Muslims," he remarked, "and the Jews are not happy to see Muslims progress. . . . If viewed from Palestine, the Jews have robbed Palestinians of everything, but they cannot do this in Malaysia, so they do this." Millions of Malaysians probably believe the accusation to this day.

Or consider the earthquake in Manzanillo, Mexico, in October 1995. Thousands of homes, offices, hotels, and factories collapsed as the shock waves rolled through cities and villages. Dozens of people were killed, and tens of thousands were left homeless. It was a tragedy for which the Mexican people, already suffering from a financial recession, were unprepared. Their reaction? Many chose to consider the Jews responsible. "Jews control the building trades," they whispered. "They built the houses cheaply. No wonder they fell down." Millions still believe the slander.

So the age-old tradition of blaming the Jews for any and every misfortune has never completely vanished. But my great fear—that this tradition would spring to life in the wake of some tragedy—had never materialized. On September 11, it happened.

Soon after the horror of the terrorist attacks, the latest version of the Big Lie began to circulate—the weird claim that Jews who worked in lower Manhattan were warned in advance to stay away from the World Trade Center on September 11. The idea is that the state of Israel, or Jews in general, were somehow complicit in the terror attacks. (In reality, of course, hundreds of Jewish Americans, and some Israelis, were among the terror victims, along with members of almost every other ethnic and national group.)

We all know that extreme, even delusional points of view can be freely disseminated on the Internet. What's more disturbing is when these bizarre lies are given broader currency by public figures who ought to know better, such as the well-known writer Amiri Baraka, who in his position as poet laureate of the state of New Jersey wrote a long diatribe about September 11 that included the "Jews were warned" claim.

When we at ADL first heard of the charge that Jews were responsible for the September 11 attacks on the World Trade Center and Pentagon, most of us dismissed it: "That's so outrageous, even the hard-core anti-Semites won't believe it." But it didn't take long for us to realize that it was not a joking matter.

Today you can travel the Arab world, Asia, and Europe and read in newspapers and hear on radio and TV the big hideous lie that has become, for countless millions, a truth. According to a Gallup poll released in March 2002, 61 percent of nearly 10,000 Muslims surveyed said they believed Arabs were not responsible for the September 11 attacks, while in a separate poll, 48 percent of Pakistanis specifically blamed the Jews. Ministers of Arab governments have bought into the Big Lie and endorsed it publicly. Syria's ambassador to Tehran, Tartky Muhammad Sager, claimed that Syria has "documented evidence" of Israel's involvement.

"Zionists pursued certain goals by conducting the attacks," he said. Saudi interior minister Prince Nayef Ibn Abd Al-Aziz told the Kuwaiti newspaper *Al Siyasa,* "We put big question marks and ask who committed the events of September 11 and who bene-fited from them. . . . I think they [the Zionists] are behind these events." Now the lie is being taught in schools in the Arab world, repeated in print and in conversations all over Europe.

Thus September 11 has helped ratchet up the anti-Semitic atmosphere around the world. So, I suspect, has the economic instability of the new century. As history shows, when the world is troubled, when people are frightened, when their leaders appear to be ineffectual against terrifying dangers, then scapegoating and hatred become tempting psychological strategies for combating the sense of helplessness and despair. And sad history shows us that, time and again, the Jews have been chosen as the targets.

It has even happened in the United States, the country above all where Jews have found a haven of tolerance and safety. During the uneasy early years of World War II, American isolationists (including Nazi sympathizers) blamed Jews for the involvement of the United States in the war; some mocked President Franklin Roosevelt as "Rosenfeld" and tried their best to undermine the war effort. During the Arab oil embargo of the 1970s, a few propagandists tried to blame the Jews for the disruption of the American and world economies. In 1991, when President Bush mobilized the world to repel the Iraqi invaders from Kuwait, right-wing politician Pat Buchanan blamed the conflict on Israel and its "amen corner" in America.

Now, in 2003, the same kinds of whispers are resurfacing. Buchanan's magazine, the *American Conservative,* loses no opportu-nity to list the members of the supposed neoconservative clique that was behind President Bush's war in Iraq. Most of the names cited are Jewish, and their sinister hidden motive is Zionism:

We charge that a cabal of polemicists and public officials seek to ensnare our country in a series of wars that are not in America's interests. We charge them with colluding with Israel to ignite those wars and destroy the Oslo Accords. We charge them with deliberately damaging U.S. relations with every state in the Arab world that defies Israel or supports the Palestinian people's right to a homeland of their own. We charge that they have alienated friends and allies all over the Islamic and Western world through their arrogance, hubris, and bellicosity. . . .

Cui bono? For whose benefit these endless wars in a region that holds nothing vital to America save oil, which the Arabs must sell us to survive? Who would benefit from a war of civilizations between the West and Islam?

Answer: one nation, one leader, one party. Israel, Sharon, Likud.

The facts belie Buchanan's paranoid fantasy. For example, a March 2003 poll by the Pew Research Center for the People and the Press found that while 62 percent of all Americans supported Bush's plan for a war in Iraq, only 52 percent of Jews supported it. (Among evangelical Christians, support for the war was 72 percent.)

Nonetheless, Buchanan's charge has been echoed from some quarters on the left. Democratic Congressman James P. Moran of Virginia declared at a public forum on March 3, 2003, that Jewish influence was behind the decision to invade Iraq. He later apologized but soon compounded the problem by saying at a party meeting that the American Israel Public Affairs Committee (AIPAC) had begun organizing against him and will "direct a campaign against me and take over the campaign of a Democratic opponent." Moran's comments echo familiar anti-Semitic slanders: that Jews secretly control non-Jewish political leaders and that they work together behind the scenes to direct policy to benefit "their kind."

Thankfully, mainstream opinion in the United States has

rejected these attempts to blame Jews for global crises. Yet in each case a vocal minority found the anti-Semitic lure attractive. And imagine what might happen in the anxiety of Americans were to be ratcheted up a notch or two—by another terrorist attack, for example, or by a prolonged and bloody war in the Middle East. Would there be further attempts by bigots to blame the Jews? Of course. Would those attempts fall on deaf ears? Not necessarily.

This is why the latent anti-Semitism that lingers below the surface in too many Americans is a matter for serious anxiety. As we've seen, the ADL anti-Semitism index reveals that one out of three Americans believes that American Jews are more loyal to Israel than to the United States. In the best of times, this would be troubling; it implies that tens of millions of Americans are prone to see Jews as potential traitors to their own country—if the interests of Israel and the United States should ever come into serious conflict. In times of crisis, like those we are now living through, this belief poses a potentially dangerous threat to the security of American Jews.

Jews are especially vulnerable today because the war on terror is inevitably linked to the Middle East conflict. Shocked by the deadly attacks on the West launched by Muslim fundamentalists, many Americans are asking, "Why do they hate us so?" There are many answers. It's clear that American culture, along with the economic and social influence of our country, is abhorrent to the extreme fundamentalists among Muslims. They reject our religious liberty, the freedom and opportunities enjoyed by American women, our sexual and cultural liberalism, and so on. But another part of the answer—and one that may seem more acceptable and understandable to the average American—is the notion that the Arab world hates America because of its support of Israel. Anti-Semites would be happy to use this idea as a wedge to separate American Jews and non-Jews.

Thus I worry that if there are more attacks, Americans—frightened, angry, bewildered—will begin to look for a scapegoat and that Jews will be the most obvious and likely target.

Do I fear a pogrom or a Holocaust in the United States? Can I envision a scenario in which hundreds of thousands of American Jews are expelled from their homes or rounded up in concentration camps? Probably not. I believe that Jews are too deeply integrated into the American social fabric—and in many cases too widely admired by their fellow Americans—for them to be readily singled out and victimized as they were in Nazi Germany. Furthermore, I believe that the American traditions of tolerance and freedom are too deeply ingrained in most Americans for this to happen (although, as the internment of Japanese Americans during World War II reminds us, these traditions may break down in times of national crisis).

So I don't literally fear an American Holocaust—and I pray that I will never have reason to change that opinion. But I can easily imagine terrible consequences, short of mass murder, that could be brought about by a new outbreak of anti-Semitism in America.

There might be political efforts to force Jews to renounce their support of Israel. (Remember the "divided loyalty" charge that anti-Semites are quick to level.) I can imagine a scenario in which, at a time of global crisis, Jewish politicians and community leaders who speak out in support of Israel are isolated, attacked, and accused of treachery.

Jew-baiting and more genteel forms of prejudice might creep back into acceptability in the media and among mainstream Americans. Jews might become social and political pariahs, as they were in large swaths of American society only a generation or two ago. Businesses might come under pressure to remove Jews from positions of power and influence (as they are already under pressure in some circles to stop doing business with the state of Israel).

Ultimately, Jews both within the United States and around the world might become increasingly isolated and vulnerable to attacks by hard-core anti-Semites. Very quickly, the actual survival of the Jewish people might once again be at risk.

SINCE SEPTEMBER 2001 we've been told over and over again that the world has changed. But some of the trends of the past two years are depressingly, frighteningly familiar. The turmoil gripping the world has been seized by bigots everywhere as a new pretext for pursuing their age-old objective: to spread hatred for the Jewish people and encourage violence against them.

So I am worried about what the future will bring. What if there is another calamity as terrible as the attacks of September 11? What if the next crisis demands an even greater degree of courage and resolution on the part of the world to stand up against terrorism and hatred, in defense of tolerance and liberty?

Perhaps the best way to prepare for the dangerous challenges of the future is to understand the failures and successes of the past. With that in mind, let's turn to the broader global history of anti-Semitism and how it has affected the lives of millions of people, Jews and non-Jews alike, over the past centuries.

CHAPTER 2

Jewish in a Hostile World:
Living with the Ongoing Battle
Against Anti-Semitism

FOR THE PAST FIFTEEN YEARS, as national director of the Anti-Defamation League, I have been a leader and a spokesperson in the ongoing struggle against all forms of bigotry and hatred, particularly hatred directed at the Jewish people. In that role I am frequently quoted in newspapers and magazines and in the electronic media. Whenever anti-Semitism is in the news—when a politician or celebrity is accused of insensitive remarks, for example, or when a synagogue is targeted in a bias attack—reporters usually call me for comment.

As a result, some people assume that I like nothing better than looking for signs of anti-Semitism under every rock. I'm sure this is one of the reasons people who don't know me well sometimes ask, "Why are you so obsessed with anti-Semitism? Aren't Jews generally well accepted in American society? Why focus on the handful of cranks and bigots who dislike them?"

Most of the people who ask this question are well-meaning. (A few are not; in fact, a few are consciously promoting the nasty stereotype of Jews as self-centered and cliquish, which is anti-Semitic in itself.) For their benefit, I want to explain why anti-Semitism is important enough for me to devote my life to it—and why the problem is a matter of such special urgency today.

To begin with, what is anti-Semitism?

Scholars who have studied the subject have developed many definitions. They have distinguished various threads and forms of anti-Semitism, such as theological anti-Semitism (driven by anger over Jewish unwillingness to accept Christian doctrines), racial anti-Semitism (regarding the Jews as inherently evil on account of some sort of "corruption in the blood"), political and social anti-Semitism (focusing on the supposedly evil influence of Jews in government, business, and culture), and scientific anti-Semitism (using trumped-up Darwinist ideas to demonstrate the biological inferiority of the Jews).

It's important that we understand and combat each of these forms of anti-Semitism. But, in truth, defining anti-Semitism and recognizing it where it exists is not complicated. Neither is it an exact science. Anti-Semitism involves human emotions, attitudes, and feelings and therefore is not subject to precise measurement. I sometimes am forced back to the old line, "If it looks like a duck, walks like a duck, and quacks like a duck, it probably *is* a duck."

Therefore, I prefer to define the syndrome of anti-Semitism in a simpler, more generic way. In the book *Living with Antisemitism: Modern Jewish Responses,* edited by Jehuda Reinhartz, Ben Halpern offers a least-common-denominator definition that says, "Antisemitism is a hostile attitude toward the Jews that has become institutional and traditional." An even simpler definition is provided in *The Encyclopedia of the Jewish Religion,* edited by R. J. Zwi Werblowsky and Geoffrey Wigoder: "Hatred of, and hostility to, the Jews." That definition works just fine in practice.

On a metaphorical level, I think of anti-Semitism as a disease. Like many diseases, it spreads from person to person. It can be inherited—not genetically, of course, but through the malign impact of a bigoted adult on his or her children and grandchildren. It can lie dormant within an individual, showing symptoms only in times of stress. And at times when a community is vulnerable, it can spread rapidly, causing an outbreak that is equivalent to an epidemic.

Anti-Semitism also resembles a disease in being fundamentally irrational. Most false beliefs are susceptible to reason and evidence. If someone sincerely believes that the Earth is flat or that sickness is caused by evil spirits, you may be able to change those beliefs by providing scientific evidence to the contrary. That's why few people in the developed world hold these false beliefs today. But because anti-Semitism is irrational, the facts are basically irrelevant. If someone calls me "a dirty Jew," showing that I have showered today will have no impact on their belief.

That's one of the reasons I and other spokespeople in the battle against anti-Semitism are generally reluctant to enter into public debates with avowed bigots—the Louis Farrakhans and David Dukes of the world. It's certainly not because we lack faith in the merits of our factual arguments. But we know—and have found through repeated experiences—that those who hate Jews do so not because of factual evidence but in spite of it. Thus dealing with the issue from a factual point of view is unlikely to change any minds. Instead, it may simply give undeserved credibility to the anti-Semites by setting up an apparent equivalency between the "two sides" in the "controversy" over anti-Semitism.

No, anti-Semitism isn't essentially a false belief (although anti-

Semites believe many untruths). Rather, it's a spiritual and psychological illness that must be combated through many means, from communication and education to social pressure and the force of the law.

A question that sometimes arises is, can a Semite be anti-Semitic? In ethnological terms, Arabs are a Semitic people. Yet as the word has come to be used, *anti-Semitism* really means "hatred of Jews"—and some in the Arab world are making it all too clear that a Semite can indeed hate Jews. So despite the etymology of the word, anti-Semitism can be found in virtually every ethnic group.

And yes, even a Jew can be anti-Semitic. I'm not referring here to so-called self-hating Jews. To me, a self-hating Jew is someone who is apologetic for being Jewish, who bends over backward to hide any signs of Jewishness and tries to distort his personality to better blend into non-Jewish society. The classic self-hating Jew changes his name, stays away from the synagogue, avoids using words and phrases he thinks sound Jewish, and maybe even has plastic surgery to alter his appearance. I find this phenomenon sad, even tragic—but I don't consider it anti-Semitism.

But there are Jews who are really anti-Semitic. Like other anti-Semites, they hate Jews and would harm them if possible. The 2002 movie *The Believer,* written and directed by Henry Bean, was loosely based on the true story of a Ku Klux Klan member who committed suicide after being "exposed" as a Jew. In the movie, the character is a Jewish youth who becomes a skinhead and a neo-Nazi.

Another example is Bobby Fischer, the great chess champion whose mother is Jewish but who has become a rabid anti-Semite. Fischer calls the Jews a "filthy, lying bastard people" who seek world domination through (among other schemes) the sale of junk food: he singles out William Rosenberg, founder of Dunkin' Donuts, as a special villain. Fischer also believes that millions of dollars' worth of chess memorabilia has been stolen from him in a secret plot involving the Rothschilds and former President Bill Clinton—who is, by the way, a secret Jew.

The sad truth is that nobody is immune from hating. You can even hate your own people—self-destructive as that may be. As the Fischer case suggests, in extreme cases, anti-Semitism can be closely akin to insanity.

THE ORIGINS OF ANTI-SEMITISM

Anti-Semitism is perhaps the most enduring form of group hatred known to human history. Although it did not assume its fullest and most destructive form until Christian times, we can see a foreshadowing of anti-Semitism in earlier periods. For example, consider this passage from the biblical book of Esther, in which Haman, an advisor to the Persian king, tries to orchestrate what amounts to an ancient version of a pogrom:

> And Haman said unto King Ahasuerus: "There is a certain people scattered abroad and dispersed among the peoples in all the provinces of thy kingdom; and their laws are diverse from those of every people; neither keep they the king's laws; therefore it prof-iteth not the king to suffer them. If it please the king, let it be written that they be destroyed; and I will pay ten thousand talents of silver into the hands of those that have the charge of the king's business, to bring it into the king's treasuries." (Esther 3:8–9)

Here are many of the familiar elements of classic anti-Semitism: the distrust of Jews as an alien people, dispersed among those of other nations yet keeping their own set of rules and customs; the sense that these aliens are disloyal and not to be trusted; and the readiness to regard the Jews as dispensable. (The rest of the biblical book, of course, recounts how the heroic queen Esther saved the Jews from the pogrom, a deliverance we still celebrate in the holiday of Purim.)

In the pagan world of the Middle East and the Mediterranean, the monotheism of the Jews and their adherence to a strict code of moral, ethical, and devotional conduct set them apart from other peoples in ways that led, perhaps inevitably, to conflict and suspicion. In the great cities of the Greek and Roman empires, for example, people of many nations mingled freely, and temples to gods from various cultures could be found side by side. Generally speaking, these varied cults lived in an atmosphere of mutual acceptance, the unspoken assumption being that "your god is yours, my god is mine, and both are equally real." Often all citizens were expected to pay taxes to support the upkeep of pagan temples, including those in which they did not personally worship.

Only the Jews took a different attitude. Divine revelation as recounted in Hebrew scripture had taught them that there was only one God; they refused to acknowledge the existence of any other gods or to pay tribute to them, financially or otherwise. Combined with the distinctive customs and religious practices of the Jews—their dietary laws, for example, as well as their reluctance to intermarry—this insistence on the exclusive authority of their own faith made the Jews appear to be a people apart, who considered themselves better than their neighbors. As a result, they were subject to distrust and sometimes hatred.

With the emergence of Christianity, anti-Semitism began to take on its tragic modern proportions. Of course, the founder of the Christian faith, Jesus of Nazareth, was himself a Jew, as were his earliest followers. In fact, it's difficult to overstate the dependence of early Christianity on Judaism. Any knowledgeable Jew who reads the Christian Gospels is struck by the extent to which the teachings of Jesus echo Jewish tradition; in fact, a large proportion of Jesus' utterances as recorded in the Christian Bible are actually quotations, free or exact, from the Hebrew scriptures. It's clear that Jesus was immersed in Jewish teaching and regarded himself as another in the line of Jewish sages and prophets, not as the founder of a new faith.

However, after Jesus' death (and, in Christian teaching, his resurrection and ascent to heaven), the community his followers established began to develop in unexpected ways. Although some of the Jews of ancient Palestine joined the new sect devoted to Jesus, most did not. Jesus' followers began to preach and teach among the gentiles, and they gradually spread the Christian teachings in Greece, Italy, Asia Minor, and elsewhere in the Mediterranean world. Soon the majority of Christians came from non-Jewish ethnic and cultural backgrounds, and the church gradually shed its original identity as a branch of Judaism.

The Christian Bible itself reflects this process. For example, chapter 11 of the Acts of the Apostles recounts a vision supposedly experienced by Peter, the preeminent leader of the early church, in which a divine messenger releases the Christians from the obligation to follow Jewish dietary laws, thereby resolving a long-standing conflict between Jewish and gentile Christians. Similarly, several of Paul's epistles offer advice on reducing conflicts between Jewish and gentile factions in the early Christian church communities.

In time, of course, Christianity became the dominant faith of Europe and the Mediterranean world. Its political ascendancy was ratified in the fourth century C.E. with the conversion of the Roman emperor Constantine, who made Christianity the official religion of the empire. As a result, the same imperial power that had previously been used to suppress Christianity (as well as Judaism) was now set the task of converting the entire population of Europe to Christianity—willingly if possible, by force if necessary.

For the next millennium the Jews of Europe occupied a peculiar position. Those who refused to convert to Christianity became a tiny religious minority in a culture where church and state were practically one. Having failed to accept the only faith that the broader society deemed legitimate, they were themselves rejected, the ultimate outsiders. Yet it was always understood that Christianity was in some sense a descendant of Judaism. The roots of Jesus' teaching in Jewish scripture were too obvious to ignore, and the Hebrew Bible itself was incorporated into the Christian Bible. How would the newly empowered Christians reconcile this duality—that while Christianity was supposed to have a monopoly on religious truth, the Jews were, spiritually speaking, their elder siblings?

In response to this contradiction, Christianity developed a set of doctrines that proved to be deadly. From the fifth century onward, Christian teachers increasingly spoke of the Jews as having been rejected by God for their obstinate refusal to accept the Savior he had sent on their behalf, Jesus. Having accepted the divine mandate that the Jews had refused, the Christian community had now become "the new Israel," a people chosen by God to replace the faithless Israelites.

Worse still, the Jews were held liable for Jesus' death by crucifixion. This accusation of deicide—God-murder—was first explicitly lodged against the Jews by Bishop Melito of Sardis as far back as the second century C.E. The falsehood of this charge has long been clear to unbiased inquirers. Over the last century a growing preponderance of evidence and scholarly study has demonstrated that the execution of Jesus was instigated primarily by the Roman authorities who ruled Palestine in the first century C.E., not by the Jewish people. And the anti-Jewish rhetoric that mars several books of the Christian New Testament has been shown to reflect not historical fact but the rivalry at the time the books were written between Jews who followed Jesus and those who did not.

Nonetheless, versions of the Gospel narratives that emphasized Jewish guilt (rather than the responsibility of the Roman imperial authorities who actually imposed and carried out the death sentence) were included in the Christian canon. As a result, with every annual reading or reenactment of the story of the death of Jesus in Christian churches, millions of Christians imbibed the notion that the Jews had been guilty of the worst crime in history. Into our own time, the deicide libel has been used to justify hatred of Jews and violence against them, including from Christian pulpits. Through the centuries these denunciations have led to countless outbreaks of violence against Jews, including murderous pogroms, a bitterly ironic betrayal of the legacy of the man Christians revere as the Prince of Peace.

THE EVOLUTION OF CHRISTIAN ANTI-SEMITISM

For almost two thousand years Christian teachings drove the spread of anti-Semitism throughout Europe and beyond. (As we'll see, the current explosion of anti-Semitism in the Muslim Middle East is fueled largely by myths and doctrines that originated in Europe.) The story of Christian anti-Semitism is a long, complicated, and tragic one. Scholars such as the late Dr. James Parkes have traced a direct line from ancient Christian teachings on Jews and Judaism to the death camps of Hitler. This book is not the place for a comprehensive account of this sad history. But some of the darkest moments in that history must at least be mentioned.

They include, for example, the horrific late-fourth-century sermons of St. John Chrysostom, bishop of Antioch, in which he condemned the "heretical" teachings of Judaism and virtually ordered Christians to launch violent attacks against the Jews, which many soon did.

Centuries later, thousands of northern European Jews were killed by soldiers of the First Crusade, in 1096, and thousands more were killed by crusaders during the subsequent centuries.

From the same time period—northern Europe in the twelfth century—we date the invention of the notorious "blood libel," the accusation that Jews murder Christian children in order to use their blood in satanic rituals. Officially, the libel was disavowed by the church, yet, as James Carroll notes in his magnificent book *Constantine's Sword,* many of the supposed victims of Jewish ritual murder, including William of Norwich, were made into saints by the church, seemingly substantiating the myth.

Given the virtual union between civil and religious authorities throughout the medieval period, it was inevitable that anti-Jewish doctrines should ultimately be reflected in political and economic restrictions on Jews. The medieval repression of the Jews throughout Christendom reached a climax in the Fourth Lateran Council of 1215, which for the first time prescribed distinctive dress for the Jews of Europe (so as to isolate and stigmatize them), banned them from public office, and even subjected them to a special tax to be paid to Christian clergy.

Later manifestations of Christian anti-Semitism include the blaming of Jewish saboteurs for the deliberate spreading of the Black Plague, which decimated Europe in the fourteenth century; the torture and killing of Jews (and other "heretics") by the Inquisition under Torquemada in Spain in the fifteenth century as well as by other inquisitors before and after him, followed by the expulsion of the Jews from Spain in 1492, and the papal bull *Cum Nimis Absurdum,* issued by Pope Paul IV in 1555, which barred Jews from owning real estate, attending Christian universities, hiring Christian servants, and engaging freely in commerce.

Until early modern times, anti-Semitism was driven primarily by religious bigotry. In fact, the avowed purpose of most anti-Semitic activity in this period was to induce the Jews to become Christians (or, at least, to prevent the spread of "false" Jewish doctrine from Jews to Christians). And there was a degree of sincerity in this belief, however misguided; the churches generally treated Jews who converted to Christianity as full-fledged Christians, with all the rights and privileges enjoyed by anyone who had been a lifelong member of the faith.

Consequently, one might imagine that the waning of the influence of institutional religion in the eighteenth and nineteenth centuries would mean a decline in anti-Semitism as well. Unfortunately, this didn't happen. Instead, as religion gave way to science, the rationale for anti-Semitism gradually morphed from a religious one to a pseudoscientific one based on doctrines of race.

The first warning signs of this new form of anti-Semitism emerged in the prescientific era, with the treatment of Jewish converts in fifteenth- and sixteenth-century Spain. No longer were converts treated as full-fledged Christians. As described by scholar George M. Frederickson in his book *Racism: A Short History*: "*Conversos* were identified and discriminated against because of the belief held by some Christians that the impurity of their blood made them incapable of experiencing a true conversion. . . . Under the doctrine of *limpieza de sangre* (purity of blood), they could still become victims of a form of discrimination that appears to have been more racial than religious."

This sinister doctrinal innovation quickly led to cruel consequences. By 1449 converted Jews were excluded from public office in Toledo. In 1492 the Jews were expelled from Spain. And by 1550 many institutions and local governments in Spain enacted "blood purity" laws, regulating who could or could not become priests, monks, conquistadors, or missionaries in order to exclude those with any taint of Jewish ancestry.

The Spanish experience presaged the emergence of full-fledged modern racism throughout Europe over the next few centuries. It may seem strange to refer to racism as modern, but as Frederickson and other specialists have shown, traditional forms of xenophobia and prejudice were qualitatively different from the racism that emerged in Europe in the early modern period. The

ancient Greek, Roman, or medieval human may have disliked foreigners or those of a different religion or skin color than himself.
But unlike the modern racist, he had no systemic worldview in
which humankind was subdivided into a number of distinct and
heritable racial categories, some considered superior and other
inferior.

This racist concept did not arise and become widespread until
the sixteenth and seventeenth centuries. Its development appears
to have been driven in part by the new encounters between
European and non-European peoples through exploration and
colonization in the Americas and Africa. Greed and lust for
power undoubtedly contributed to the appeal of racism. The
European explorers and conquerors who "discovered" the rich
new lands beyond the seas were eager to wrest control of them
from the (mostly) dark-skinned peoples who lived there. But as
Christians, they felt the need to morally justify their imperialist
designs. A doctrine that radically differentiated "white" people
from "Black" and other "colored" races, relegating the latter to a
status little higher than that of animals, ideally served this purpose.

Racism took on a pseudoscientific form in the nineteenth century, with the emergence of the doctrine of social Darwinism.
According to this belief (which no reputable geneticist, anthro-
pologist, or biologist holds today), the varying races of human
beings were regarded as analogous to differing species, some of
which were supposed to be higher or more advanced than others.
All of these races were engaged in a struggle for survival, which
would inevitably lead to the triumph of the "fittest" specimens.
Under this doctrine, Blacks, Asians, and other non-Europeans
were stigmatized as inherently inferior and therefore worthy of
subjugation or even extermination by white Europeans. The Jews
were soon fitted into this framework as a Semitic "race," inferior
to the Nordic or Aryan peoples among whom they lived.

If anything, the new racially based strain of anti-Semitism was even deadlier than the purely religious strain. If Judaism was a race whose evil nature was somehow inherent in the blood or the genetic inheritance of Jews, then conversion, education, or other "advances" could make no real difference in the status of a Jew. In fact, allowing Jews to integrate into the mainstream could only spell disaster, since the Jewish "taint" would gradually spread and, through its inherent corruption, bring about the downfall of society.

The racist and religious forms of anti-Semitism merged in Pope Pius IX's *Syllabus of Errors* (1864), which attacked Jews as carriers of the "infections" of modernism, liberalism, socialism, and other supposed evils of contemporary life. And of course, it reached its climax of power and destructiveness in the life and work of Adolf Hitler, the Nazi doctrine of racial purity, and the Holocaust, which Hitler regarded as the "final solution" of Europe's Jewish problem.

Why did Christians allow themselves to be consumed by anti-Semitism for so long? What psychological forces drove Christians for twenty centuries to single out a relatively small, relatively helpless minority in their midst as the victims of invective, hatred, and murderous violence?

There are several possible answers, though none seems definitive. Perhaps, as the American novelist and social critic Mark Twain supposed, some Christians are consumed with envy over the apparent success of Jews in so many diverse fields of endeavor— or even over their biblical status as God's chosen people. Perhaps Judaism, as the elder brother of Christianity and the ancient source from which the younger faith sprang, is the unfortunate victim of an extreme case of sibling rivalry—the innocent Jewish Abel to the Christian Cain.

Or perhaps some evil element in human nature demands a scapegoat on whom we can act out all the frustrations and dissatisfactions of life, and perhaps the Jews simply represent a safe and convenient target for such scapegoating. If this explanation is correct, the phenomenon of anti-Semitism may seem even more troubling; if it is driven by so elemental a need, perhaps there is in the end no hope of ever extirpating it completely.

Novelist Anne Roiphe speculates that anti-Semitism, like all forms of racism and bigotry, may ultimately derive from childhood fears and traumas:

Antisemitic literature is rife with descriptions of disgust. Jews are said to smell bad, their clothes offend, their breath is bad, they are said to be repulsive and to kill and eat little Christian babies. The Jew then is in society what the stranger is to the nursery: the feared one, the person who rips apart the dream of permanent safety, of mother's arms eternally enclosing and protecting. Why, one wonders, should such an infantile thought persist in the adult and become transformed into the rules of the country club that doesn't permit Jews or the university that holds quotas or the Nuremberg laws that forbade intermarriage? Why do the Hutus consider the Tutsis the stranger and vice versa? Why do the Serbs and Croats and the Muslims use the most minor of differences to draw bloody lines across their maps?

It must be that the human mind matures but holds on to its oldest of traumas, its most primitive of fears; its first thoughts lie just beneath its second and third and shape them, direct them, attach to the most complicated. Prejudice is then an echo of a child's scream, Where is my mother and who are you?

Great artists, philosophers, prophets, and saints have always wondered about the origins of good and evil. In the end, they are mysterious. So it is with anti-Semitism. We can easily trace its historic roots and its development in theology, politics, and culture, but its ultimate source in the recesses of the human heart remains elusive.

ANTI-SEMITISM TODAY

In the immediate aftermath of World War II, the universal revulsion at the horrors of the Holocaust were widely viewed as spelling the effective end of anti-Semitism. By revealing the ultimate destructiveness and insanity of anti-Semitism, Hitler was supposed to have rendered the doctrine abhorrent to any civilized society. Sadly, it isn't true. Today, sixty years after the Holocaust, anti-Semitism is alive and well.

At ADL, we track the prevalence of anti-Semitism in a number of ways. We monitor anti-Semitic hate crimes and bias incidents as well as statements and actions by public figures, political leaders, and the media that support anti-Semitic views. We also conduct periodic surveys designed to identify anti-Semitic attitudes and beliefs, using (among other measurement tools) the anti-Semitism index described in the previous chapter.

The evidence we've gathered demonstrates that anti-Semitic attitudes continue to exist in the United States and around the world, despite everything—despite the lessons of the Holocaust, despite decades of outreach efforts in schools and mass media, despite the commitment of virtually all responsible politicians and political groups to opposing and eradicating anti-Semitism. And as we've already discussed, the last few years have brought frightening signs of a resurgence of this age-old form of hatred—in Europe, in the Middle East, in South America, and around the world.

The anti-Jewish beliefs uncovered by our surveys are latent attitudes—ones that are generally not acted upon or even openly discussed. Most people don't get up in the morning and ask themselves, "Now, how can I stick it to the Jews today?" (There are some who do! But in most countries, they are a relatively isolated and small minority, though not without influence.)

In some ways, we have made genuine progress in the war on

hatred, particularly in the United States. In America today, most Jews don't face the same kinds of barriers to acceptance that they faced a century or even a generation ago. Thanks to widespread changes in social attitudes, a growing atmosphere of tolerance, and forty years' worth of civil rights legislation ensuring the rights of all Americans, there are few occupations, schools, clubs, and communities where Jews are not accepted. Jews hold prominent roles in business, they have been elected and appointed to important positions in government, and Jews can attend any college or university in the country.

Perhaps the most positive sign I've seen in American attitudes toward Jews was the reaction to the vice presidential candidacy of Senator Joe Lieberman in 2000. When Democratic nominee Al Gore selected this observant Jew as his running mate, I and many of my fellow Jews were elated—but also a little worried. Would the emergence of a Jew in such a prominent political role stir up the old hatred? Would Lieberman's religion make him a target for overtly or subtly bigoted attacks like those to which the Roman Catholic John F. Kennedy was subjected forty years earlier? And if the Gore-Lieberman ticket faltered, would enemies of the Jews seize the opportunity to blame Lieberman for its failure?

None of these things happened. Not only did Lieberman's religion *not* become a campaign issue, most observers agreed that it was a significant political asset to the Gore-Lieberman team. (Of course, we'll never know for certain how many voters failed to pull the Democratic lever because of Lieberman's place on the ticket.) And even when the Democrats were defeated in one of the most contentious and controversial elections in history, no one accused "the Jew" of having sabotaged the ticket. In many ways, the 2000 election was a watershed for American Jews, a sign that the political and cultural attitudes of Americans have reached a significant level of maturity. Although latent anti-Semitism remains widespread, it doesn't necessarily find expression—not even in a situation, such as

the Lieberman candidacy, that might well have evoked it a generation ago.

THE ERUPTION OF ANTI-SEMITISM IN TIMES OF CRISIS

Given that there is progress to celebrate, why do I worry about
latent attitudes of hostility toward Jews—and more important,
why should others, Jews and non-Jews alike, share my concern?
There are several answers.

First, experience shows us that anti-Semitism is like the canary
in the coal mine. In the days before electronic devices for measur-
ing air quality, miners would bring canaries underground to serve
as early-warning devices. When the little birds became sick, the
miners knew that deadly gases were accumulating in the tunnels
and it was time to flee to the surface. In the same way, Jews are
usually among the first to suffer when bigots, tyrants, and oppres-
sors rise to power and prominence in a society.

In a sense, the wise and good Albert Einstein was mistaken
when he described anti-Semitism as "the shadow of the Jewish
people." A shadow can't exist in the absence of the body that casts
it. Yet in our lifetime we have seen anti-Semitism in countries
with few or no Jews, including Japan, Malaysia, and Poland. Thus
those who are prone to hatred don't become anti-Semitic in reaction to the presence of Jews. Rather, it's as if they seek out Jews
because they are such a convenient target for abuse.

Thus how a society deals with the Jews is a sensitive measure of
its commitment to democracy. When anti-Semitism is allowed or
even encouraged to flourish, freedom is in danger—not just for
Jews, but ultimately for every member of a social, ethnic, religious, political, or sexual minority. Before Hitler was done, he had
targeted not only Jews but trade unionists, gypsies, homosexuals,
the mentally ill, and many others. And failure to recognize and

pay attention to the danger of anti-Semitism is itself a worrisome first step toward totalitarianism and violence. Latent anti-Semitic attitudes must be monitored and responded to because they represent early warning signs about the health of a democracy.

That leads to my second point about the importance of latent attitudes. Two thousand years of history demonstrate that anti-Semitic attitudes can lead to actions—and have done so all too often. So latent attitudes are likely to be expressed, in words and actions, in times of crisis, whether personal, communal, or national. And when that happens, the consequences can be serious.

Here's an example of what I mean by a personal moment of crisis. Air Force General George S. Brown served as chairman of the joint chiefs of staff under three American presidents (Nixon, Ford, and Carter). By all accounts, Brown was an effective military leader. He was educated at West Point. During World War II he won the Distinguished Service Cross for his heroism in leading bombing raids over Nazi-run oil refineries in Romania. He later served in Korea and Vietnam, rose through the ranks, and became chairman of the joint chiefs in July 1974 (just days before the historic resignation of President Nixon).

I don't doubt that General Brown was in many ways a good man. But he was infected with the virus of anti-Semitism. How do we know? It emerged in a personal moment of crisis.

One of Brown's responsibilities as chairman of the joint chiefs was to monitor and maintain the stockpiles of weapons provided to NATO forces in Europe. He took this responsibility seriously. After all, he had served during some of the tensest moments of the Cold War, when the threat of a Soviet invasion of Western Europe was perceived as very real. Therefore, in October 1974, when President Ford ordered him to ship arms to Israel on an emergency basis because of heightened tensions in the Middle East, Brown was deeply upset. Doing this, he feared, would deplete the NATO stocks. What if the Soviet Union chose this moment to attack the West? Could Europe fall because of a lack of materiel—for which Brown would be responsible?

The general could have responded to this dilemma in any number of ways. Unfortunately, the way he chose was to burst out with a classic explosion of anti-Semitic hatred. In his view, Ford was sacrificing the safety of Europe in order to appease Jewish voters. Among other comments, Brown was quoted as saying, "They own, you know, the banks in this country, the newspapers. Just look at where the Jewish money is."

When Brown's remarks became public, Jewish groups naturally responded with outrage. But as far as we know, he was never rebuked by President Ford. Although Brown later made similar remarks when speaking to the Senate Armed Services Committee, provoking further dismay, he remained in office until June 1978, just six months before he died.

In the end, neither American Jews nor the people of Israel were physically harmed by anything Brown did or said. But it doesn't take much imagination to think of circumstances in which an American military leader with such attitudes might be tempted to abandon our Israeli allies in a time of crisis. A debacle with terrible global consequences could result.

This is why I have paid such close attention to the signs of latent anti-Semitism—or of flirtation with anti-Semitism—in such prominent public figures as Pat Buchanan, Jesse Jackson, William F. Buckley Jr., and Al Sharpton. In time of personal crisis, latent attitudes become explicit, and individuals with influence and authority can do real damage if their attitudes are marred by prejudice and hatred.

On a larger scale, entire communities can experience similar flash points—crises in which latent anti-Semitic attitudes come to the fore. Sometimes these can have deadly results. Consider the horrible August 1991 crisis in Crown Heights, Brooklyn, which led to riots, street fighting between Black and Jewish crowds, and ultimately the stabbing murder of nineteen-year-old Yankel Rosenbaum, a religious student.

In Crown Heights, the flash point was a car accident. A vehicle in the motorcade of the Lubavicher Rebbe, the leader of the Lubavicher Hasidic community, struck and killed an African American youngster named Gavin Cato. A series of missteps by local and city leaders turned this tragedy into a crisis. Black residents of Crown Heights became convinced that the police and civic authorities were showing favoritism to the Hasidic community in their handling of the Cato incident. Thanks to miscommunication and indecision on the part of City Hall, police failed to control the crowds when they marched in protest, demanding harsh retribution against the driver of the car.

In the resulting turmoil, a group of Black youths spotted young Rosenbaum's distinctive Hasidic garb and shouted, "Kill the Jew!" choosing him as a Jewish sacrifice to pay for the death of Gavin Cato. Soon Rosenbaum lay dead. It was a pogrom in New York.

It took weeks for community leaders to restore calm and prevent further bloodshed. Nothing, of course, can ever restore Yankel Rosenbaum to his family. And the breach between the Black and Jewish communities of Brooklyn still has not fully healed, more than a decade later.

The blame can be laid at the feet of the latent anti-Semitic attitudes that linger in too many communities, like a low-grade fever, waiting for a moment of tragedy, misunderstanding, or conflict to erupt.

Most dangerous of all are the moments of national or even global crisis, when latent anti-Semitism has the potential to produce truly horrific results. As we've seen, these are the times when anti-Semitism becomes murderous, producing pogroms or even the ultimate nightmare of the Holocaust. And today's historical moment—the era of the *intifada,* the war on terrorism, the deepening conflict in the Middle East, and the broader clash between Islam and the West—represents perhaps the most dangerous time of crisis since the 1930s.

This is why ADL takes its mandate to monitor, challenge, and oppose anti-Semitic attitudes so seriously. It's not that we are thin-skinned or that we want everyone to love the Jews. It's because harsh experience has shown us that recognizing symptoms of anti-Semitism and stopping it before a new epidemic has a chance to explode are, for us, matters of life and death—literally.

How Jewish Life Has Been Warped by Anti-Semitism

One of the most corrosive effects of anti-Semitism is the way in which it distorts the behavior and attitudes of the Jews themselves.

Recent reports out of France detail how French Jews are reacting to the current upsurge in anti-Semitism there. Some are moving to Israel, an act that constitutes a strong affirmation of their faith. (That's the only reason a Jew would move into terrorist-besieged Israel; it can't be in search of increased personal safety.) But the other response is to become less Jewish—to stay away from the synagogue, to refrain from sending the children to Jewish schools, to avoid talking about one's faith, maybe even to change one's name.

So anti-Semitism can affect Jews in at least two contradictory ways: it can strengthen their faith or weaken it, depending on personal circumstances, knowledge, and commitment.

This is not to say that there is a silver lining to the dark cloud of anti-Semitism. No Jews want to live with their neighbors' hatred, and the Jewish community never benefits from a climate of bigotry. It's inevitably difficult to be a member of a small, beleaguered minority, and Jewish culture and religion have certainly suffered under the pressure of historic anti-Semitism.

The last two millennia have not been easy for the Jews! As columnist George Will has pointed out, if the proportion of Jews in the world today were the same as it was in the time of the Roman Empire, there would be 200 million Jews. Instead, there are only 13 million. Pogroms, hate crimes, and a Holocaust are not a healthy environment for any people.

Over the centuries, anti-Semitism also put enormous pressure on Jews to abandon their faith through assimilation and conversion. Today these factors continue to play a powerful role in the decline of the Jews. The numbers are distressing. They include a relatively low birth rate among Jews, a high intermarriage rate (leading to many conversions out of Judaism or a quiet abandonment of religion), and a low rate of conversion into Judaism.

Ironically, the efforts of ADL and other organizations that combat anti-Semitism probably make it easier for some Jews to assimilate. After all, by tearing down the barriers that separate Jews from the broader society, we encourage more social and professional contacts between Jews and non-Jews, which inevitably lead to marriages and conversions. But of course this is a risk we have to take. We must believe that Judaism will survive because of its positive aspects—its cultural and spiritual richness—rather than because of walls that divide the Jews from the rest of the world. We must believe that the next generation of Jews will choose to retain their faith rather than discard it.

Nurturing the Jewish Faith in a Hostile World

However, if Jews are to make intelligent choices about the future of their faith, the first requirement is education and knowledge. You can neither accept nor reject Judaism, its history, and its values unless you've been taught about these things and have a serious understanding of what they mean.

Here is where we are falling down as a community. Too many young Jews receive only rudimentary religious training. And those who do often stop studying and practicing their faith at age thirteen, right after bar mitzvah or bat mitzvah. Throughout the crucial teenage years, when peer pressures are intense and young people are truly learning what it means to be responsible adults in a confusing and challenging world, we neglect their spiritual development and allow the highly secular mass media and educational systems to work on them unchecked. Then when they go off to college, we hope that Hillel and other Jewish campus organizations will attract them. But why should they, when our Jewish youth have seen religion being treated as unimportant in their own homes and communities?

Indirectly, anti-Semitism plays a role in our failure to nurture the faith of our young people. I think many Jews are so eager to blend into American society that they shy away from activities that seem "too Jewish," including attendance at Hebrew school and religious observances. It may appease the anti-Semites, but in the long run it spells doom for the Jewish people just as certainly as a second Holocaust.

When I discuss this problem with my Jewish friends, some of them say, "We don't believe in pushing faith on the kids. They can choose for themselves when they grow up." This attitude may be well-meaning, but I think it's a delusion. The choice being referred to is whether to accept or reject Judaism—but you can't intelligently accept or reject something you know nothing about. And in our struggle to assimilate and be accepted in the mainstream of American society, many Jews have given up not only the outer trimmings of religion but also the fundamental knowledge of our faith. As a result, we have lost two generations of young Jews.

The problem is even more complicated in interfaith marriages. When a Jew marries a Christian, I think the new couple ought to choose one faith or the other and raise their children consistently in that faith. I know that we will lose some Jews in this way, but I think this is fairer than trying to raise kids in two religions, as many families say they do. When you try to do justice to two sets of rich and beautiful traditions, rituals, beliefs, and value systems, you inevitably shortchange both.

I believe that Judaism will never disappear. Jewish tradition is too rich, and the human yearning for a connection to God is too strong, for our faith to vanish altogether. Many young people today are returning to their Jewish heritage, and many more will do so if we adults provide them with the opportunities to learn, share, and worship together. It's a responsibility we have not carried out as fully as we should.

I've always been zealous about the importance of transmitting our faith to the next generation. Before joining ADL, I spent a couple of summers as program director at a Jewish camp in the Midwest. One summer I constructed a three-week educational program for our oldest group of campers, ages fourteen to sixteen. It was built on excerpts from the great books of Judaism, including the Bible, the Talmud, and the Zohar (the central work of Kabbalah, the system of Jewish mysticism). We also read selections from later writings, including three or four religious books, the classic Zionist text *Der Judenstaat* by Theodor Herzl, some works of the great Yiddish authors Isaac L. Peretz and Sholem Aleichem, and a few selections of modern Hebrew literature.

The campers found the program provocative and exciting, but it stirred up a fair amount of controversy among parents and some of the other staff members. I was accused of trying to indoctrinate the young people in Orthodoxy. "Not true," I replied. "Orthodox, Conservative, Reform—whatever form of Judaism you believe in, you need to understand its roots. After all, in order to reform something, you need to know what you are reforming. Otherwise, your actions will be based in ignorance."

I still think young Jews need the kind of immersion course in Jewish teachings and tradition that I designed all those years ago. If I could, I would make Jewish day school education at all levels available and affordable for as many Jewish families as possible. This would mean building many new schools and providing scholarships for those who need them. It would be expensive— but worth every penny. Such a program could help bring about a vibrant Jewish future in America, the "golden medina" (golden land) of our people.

ISRAEL AND JEWISH CULTURE

There's another step I would take. I would send every kid who was willing to go for an experience of the state of Israel. Such a program would act as a tourniquet to stop the slow bleeding of Jewishness from our people.

Israel is important for Jews precisely because of the legacy of anti-Semitism. Everywhere else in the world, Jews are a small, sometimes tolerated, often hated minority. But not in Israel. The difference is profoundly important.

Rabbi David Hartman, founder of the Shalom Hartman Institute in Jerusalem, puts it well. He says that God has given us Israel as a place where we can experiment with Judaism as a *living* religion. That is, Israel is a land with a majority Jewish culture, where there are no phobias or hang-ups about being Jewish, where being Jewish is natural and normal and requires no apologies. In Israel we see everyday things that are rare sights in other countries: Jewish police officers, Jewish athletes, Jewish bus drivers, Jewish soldiers! For any Jew, to experience such a world is an amazing and liberating thing.

I don't mean to imply that Jews in America are being crushed under anti-Semitism today. The problem I am describing is a subtle thing. Under the best of circumstances, a member of a minority group, especially one that is scrutinized and criticized like the Jews, is always concerned, deep inside, about what the people of the majority are thinking and feeling about us. This concern distorts our feelings and behaviors in many ways. It makes some Jews say, "Perhaps we shouldn't stick together so much." It makes others say, "We shouldn't act too 'Jewish.'" Still others live in a constant state of low-level anxiety, hearing a voice inside them whispering, "Don't be too noisy, too successful, too prominent, too pushy."

I once had a conversation about being Jewish with the late Edmond Safra, the founder of Republic Bank. He said to me, "When we opened up banking operations in Latin America, I called in my managers and told them, 'I'm sure we will be successful and profitable here. But as we grow, I want you to make sure we are always the second or third largest bank in the country, never the first. As Jews, we can't afford the visibility of being the biggest.'" It's a feeling that millions of Jews harbor deep inside, often unconsciously.

For many Jews, visiting Israel helps take away this sense of insecurity. In a land where being a Jew is normal, there's nothing a Jew can't do.

ISRAEL AND JEWISH SURVIVAL

Israel, then, is crucial for the survival and health of Jewish culture. It's also important to the Jews on a still more basic level. After the Holocaust Israel is also the place of last refuge for endangered Jews, the country where Jews can go when they are rejected everywhere else.

One of the most shameful aspects of the tragedy of the Holocaust is that so many died because they had no place to go. Of course, thousands of refugees did flee to safety in America and other countries around the world, but hundreds of thousands more were rejected and denied asylum. There are many tragic episodes in this story:

- Throughout the 1930s and the war years, Canada refused to admit Jews fleeing persecution, including thousands of orphans who sought refuge there.

- In May 1939 the U.S. State Department refused entry to the SS *St. Louis* and its cargo of 930 Jewish refugees. The ship was forced to return to Europe, where most of the passengers died in the Holocaust.

- In 1944 Hungary could have bartered supplies, including trucks, in exchange for the freedom of a million Jewish refugees but passed up on the opportunity because Hungary was unwilling to absorb so many Jews.

In fact, at the Evian Conference, convened by President Roosevelt in 1938, the unwillingness of the Western nations to accept large numbers of Jewish refugees from Hitler was made into virtually official policy by the United States and its European allies. Much later, the American War Refugee Board helped a few (especially supporting the efforts of the heroic Swedish diplomat Raoul Wallenberg), but it was a case of too little, too late.

The Holocaust deepened the Jewish sense of anxiety about depending on the goodwill of others, which two millennia of tragedy had already developed. The existence of Israel, a home-land established by Jews for Jews, serves as a psychological insur-ance policy for the Jews, who are well aware that if there had been an Israel during World War II, many of the six million would have been saved.

And the idea of Israel as a land of refuge for the Jews is not just symbolic or theoretical. It has been put into practice at least twice. The first instance was in the 1970s, when, under strong pressure from the United States (via the Jackson-Vanik Amendment) and from Jews around the world, the Soviet Union, which had harshly suppressed the practice of Judaism, agreed to allow Jews to emigrate to the Jewish homeland in what came to be called Operation Moses. The second instance was Operation Solomon (1984–85), which liberated persecuted Ethiopian Jews to emigrate to Israel in similar fashion.

ADL played a role in both of these liberation movements, and we stand ready to help launch similar movements in other countries if they are ever needed. Among the nations we are watching with this concern in mind are Argentina, South Africa, Morocco, Tunisia, and Iran. In extreme circumstances, even some European countries, such as France, could see an exodus of the Jews. Without the existence of Israel, this final option for survival would be impossible.

Sometimes people accuse me of being too much of a hard-liner on issues related to the state of Israel. I plead guilty to the charge. History has proven that a strong Jewish state is essential to the long-term survival of our people. That's why I have never compromised on Israeli security, and never will.

ISRAEL AND THE CHALLENGE OF JEWISH IDENTITY

The founding of the state of Israel represented a crucial moment of hope for the Jewish people. Yet its existence also creates difficult existential challenges for Jews.

The fact that Israel is a Jewish state raises all kinds of questions. What exactly does "a Jewish state" mean? What should be the status of non-Jewish faiths in Israel? What political, social, and religious rights and privileges should non-Jews enjoy? And on an even more fundamental level, who is a Jew? And what constitutes the Judaism that Israel seeks to embody and preserve?

All of these questions of Jewish identity have been debated and discussed throughout history. Like all religious and national groups, the Jews have wrestled with issues of self-definition. But for the past two thousand years these issues were purely theoretical. Although the Jews maintained their own religious and cultural life, it was Christian theologians and rulers in Europe, and Muslim rulers in the Middle East, who decided who must wear the yellow star, who must live in the Jewish district, who would be restricted in business and social life. As a small, relatively power-less minority, the Jews had no choice but to go along.

The establishment of modern Israel in 1948 changed all that. It created the first opportunity in two millennia for Jews to deter-mine for themselves who they are and how they ought to live. Yet for historic reasons, the Jews of Israel have been unable fully to accept this challenge.

One reason is the circumstances of the founding of Israel and the early history of the state.

Not all Jews supported the notion of a Jewish state. There were many Jews and Jewish organizations who were anti-Zionist or, at best, indifferent to the Zionist cause. Some, like the American Council of Judaism, took this position in part because they feared a revival of the old double loyalty accusation so often used by anti-Semites against the Jews. Others opposed a Jewish state for political reasons; for example, believers in universal, one-world government opposed Zionism as a new version of the nationalism they abhorred as the source of conflict and war, and the Com-munists opposed it because they considered it a distraction from their real mission, the overthrow of capitalism.

Some Jews, especially some of the ultra Orthodox, opposed the creation of Israel on religious grounds, based on traditional teach-ings that say the Messiah must come *before* the nation of Israel can be reestablished. To this day, some reject the legitimacy of the current state of Israel because, in their view, only God should determine when and how a Jewish state should be created. For example, the Neturei Karta sect believes that "One of the basics of Judaism is that we are a people in exile due to Divine decree. Accordingly, we are opposed to the ideology of Zionism, a recent innovation, which seeks to force the end of exile." Members of the sect demonstrate regularly outside the Israeli consulate in New York and, oddly, voice support for the likes of Louis Farrakhan and Yasser Arafat. When the Soviet Jews were leaving Russia in the 1970s, Neturei Karta members would meet them in Vienna and offer to pay them to go to America rather than Israel.

(The Neturei Karta sect and other similar ultra-Orthodox groups differ from the mainstream Hasidic movement, which generally regards the founding of Israel as symbolic of the beginning of the redemption of the world. The existence of Israel is not only acceptable in their eyes but a positive sign that the coming of the Messiah is on the horizon.)

Thus the Jewish community was historically divided over the importance and even the desirability of the existence of Israel. And although the dream of a Jewish state was religiously motivated, biblically based, and theologically grounded, the founders of Israel were *not* deeply religious people. Theodor Herzl, the founder of modern Zionism, was an assimilated Austrian Jew, and his mission, to establish a Jewish homeland, did not necessarily focus on the historic importance of the Holy Land. At the Sixth Zionist Congress, in 1903, Herzl proposed the establishment of a Jewish state in Uganda, if that were practical. (Members of the Russian and Polish delegations left the hall in tears.) Later, when Israel was a going concern, the nation's first leaders, including David Ben-Gurion and Golda Meir, were nonpracticing Jews whose primary ideology was socialism.

Culturally, too, Israel is far from a simple or monolithic place. Although most of its founders were ethnically and culturally European, today almost half the population is descended from

Jews who fled or were expelled by the neighboring countries of Iraq, Syria, Yemen, Lebanon, Morocco, and Egypt in 1948. These immigrants have brought a strong Middle Eastern flavor to Israeli culture.

Thus the newly founded Israel was a mélange of many ethnicities, cultures, and religious and political points of view. Forging a distinct and unified nationality from this mixture would not be an easy task. But because the new nation was immediately attacked by its Arab neighbors, the leaders of Israel postponed debates about national identity, fearing that any attempt to steer Israel in a single direction might shatter the unity of the Jewish people and impede their efforts to defeat the common enemy.

Under the circumstances, issues that had a potential impact on cultural and national identity were decided on a purely pragmatic basis. For example, it was decreed that the Israeli army would observe the sabbath and prepare meals according to kosher restrictions. Why? Golda Meir, who did not keep a kosher kitchen in her own home, observed that it wouldn't hurt a nonkosher person to eat kosher food, but the reverse would be a terrible imposition. Thus observing the kosher rules would be simpler and more politically acceptable than disregarding them.

In this way a potentially delicate issue was postponed. In the process, however, religious Orthodoxy was subtly granted a political role in Israeli life that it retains to this day.

When the founders of Israel decided to postpone questions of national identity until peace could be established, they didn't expect the issues to hang fire for over fifty years. But today the sense of being besieged continues to enforce an uneasy unity among Jews both inside and outside Israel. In fact, the continuing state of war between Israel and most of its Arab neighbors has gradually strengthened the solidarity of Jewish support for Israel.

Before 1967 most Jews around the world were not Zionists; they may not have opposed the existence of Israel, but they didn't necessarily support it. That began to change in 1967, when Egyptian president Gamal Abdel Nasser gathered a coalition of Arab states in an effort to destroy Israel. Jews saw how the world failed to respond to the pleas of Israeli diplomat Abba Eban to uphold international law and force Nasser to reopen the Straits of Tiran and the Suez Canal. This crisis, and the subsequent Six-Day War, had a sobering effect on Jews the world over. For many it brought home the realization that a second Holocaust was far from impossible. As a result most Jews rallied around Israel and have remained fundamentally Zionist in orientation ever since. Their adherence to Israel has been reinforced by more recent events, especially the passage of the notorious "Zionism equals racism" resolution by the United Nations in 1975.

As long as Israel continues to be threatened with destruction, nearly all Jews will support it, and the issues of who is a Jew and how a Jewish state should be governed will remain on the back burner. After the signing of the Oslo Accords in 1993, there was a widespread perception that the Middle East was at the brink of peace. As a result the debate about Jewish identity and the future of Israel started up. But soon the unhappy realization that peace was still far away returned and the debate subsided.

Nonetheless, everyone knows that these questions must some-day be dealt with. The tension between the notion of a Jewish state and the impulse toward democracy in Israel is a real one. In the grip of this tension, we've seen Jews express and practice big-otry not far removed from the bigotry by which Jews have been victimized for centuries. We've seen Jews make bigoted remarks about Arabs (and at times victimize Arabs economically and mili-tarily), and we've seen secular Jews and religious Jews treating one another in hateful, bigoted ways. We've even seen Jewish big-otry explode into violence, as when Israeli prime minister Yitzhak Rabin was assassinated in 1995 following accusations that he was a *judenrat*—that is, a quisling. The ADL has spoken out forcefully against such words and actions, no matter who is responsible for them, and will always do so.

These intracommunal tensions have spilled over into other countries where Jews live, including the United States. I formerly belonged to an Orthodox congregation in Teaneck, New Jersey, where the struggle for survival and peace in the Middle East was a deeply felt and highly personal issue. During the period of Prime Minister Rabin's peace initiatives, the rabbi used his pulpit to criticize Rabin harshly, accusing him of selling out Israel to its enemies. At the rabbi's direction, the congregation even stopped offering the traditional prayers for the welfare of the state of Israel and its leaders.

I was deeply troubled by the intrusion of political differences into our religious observances; I felt that the rabbi's statements and actions were intolerant and fueled hatred. When my efforts to get other lay leaders in the congregation to speak up in protest failed, I wrote a public letter announcing that I was leaving the synagogue after twenty-five years there. The controversy received much more attention than I expected. My stance drew a front-page story in the *New York Times* and follow-up columns by the likes of Thomas Friedman.

For my part, I think that Israel must move toward American-style separation of religion and the state, especially as the region moves closer to peace. I understand the historical need for some restrictions on religious practices. For example, there are laws in place that limit proselytizing by Christian missionaries, which I think are appropriate under current circumstances. (Some Christian groups were even said to be offering cash payments to Jews for converting, which I find very distasteful.) But ultimately, faith can survive only as a positive good, not because the alternatives are restricted by law.

I believe in my heart that Judaism will survive, despite efforts by other faiths to seduce Jews into their flocks. And I believe that

Israel should strive to live up to its heritage as a true democracy, in which government plays no role in determining or influencing the religious practices of the people.

Disentangling church from state in Israel will not be easy. There are many points of contact, and the questions raised are not simple ones. Take the matter of the sabbath, a religiously mandated day of rest. For Jews, the sabbath is Saturday; for Muslims, it's Friday; for Christians, it's Sunday. In Israel today, Saturday is treated as the sabbath by government, business, and the schools. But what if you're not a Jew? Suppose you'd like to spend Saturday on one of the beautiful beaches near Jerusalem. The buses don't run on Saturday, and the poor can't afford taxis. Removing these restrictions would be fairer for all of Israel's people, but it would also push Israel further away from being a Jewish state.

Or take the question, who is a rabbi? Does Judaism equally embrace its Orthodox, Conservative, Reform, and Reconstructionist branches? Who should be authorized by the state to perform rituals like marriage? In Israel today the Orthodox persuasion dominates. There are a few conservative and reform synagogues and schools, but not many. In time, this must change. Ultimately, all traditions must be respected equally.

These fundamental questions about Jewish identity and the role of Israel are so troubling, in large part, because of the corrosive historic impact of anti-Semitism. Our desperate need for a secure homeland—our longing for a place of normalcy, where we can be ourselves without anxiety and fear—and our uncertainty as to what it means to be a "real" Jew have been deeply colored by twenty centuries of life under siege, surrounded by a world that distrusts and hates us. In the next chapter, we'll consider how this came to be—and how a great institution supposedly dedicated to forgiveness, redemption, and love, the Catholic Church, became the world's greatest promoter of anti-Semitic intolerance, bigotry, and hatred.

CHAPTER 3

Cradle of Hatred:
The Tragedy of Jewish-Catholic Relations

WITH TWO THOUSAND YEARS of history, a magnificent legacy of theological wisdom, a beautiful liturgy, and hundreds of millions of members around the globe, the Roman Catholic Church is obviously one of the world's great religious institutions. Catholic leaders and laypeople have performed countless acts of charity, generosity, and heroism; Catholic artists have enriched world culture with masterpieces of religiously inspired music, painting, sculpture, and literature; and Catholic historians, scholars, and philosophers have profoundly advanced the human search for truth.

All true. Yet this proud history exists side by side with a legacy of incredible shame and horror. For almost twenty centuries, during many of which the Roman Catholic Church was *the* Christian church and the primary institutional heir of the teachings of

Jesus, the church was the archenemy of the Jews—our most powerful and relentless oppressor and the world's greatest force for the dissemination of anti-Semitic beliefs and the instigation of violent acts of hatred.

THE CHURCH AND THE HOLOCAUST

This horror extends its stain to the threshold of our own times. The culminating act of the anti-Semitic litany of death, which all decent humans pray was the final act of the tragedy—the Holocaust—took place in Europe some three generations ago. Within living memory, millions of innocent Jews were slaughtered by Christians in the willing service of a psychopathic dictator. Many of the same people who operated the gas chambers worshiped in Christian churches on Sunday.

Obviously there is plenty of guilt to go around in the epic horror of the Holocaust. The creators and promulgators of the Nazi ideology, and those who eagerly supported them, bear the greatest responsibility. But for Christians and Jews of today, still struggling to make a moral accounting for the most terrible crime in human history, the question of the complicity of the church in the murder of the Jews is a living one. We must understand the truths of our history—even the most awful truths—if we hope to avoid reliving them.

For some, the chief question revolves around the guilt of Pope Pius XII, the leader of the church during the years before and during the Second World War. His actions—and his failure to act—at the time of the Holocaust are the subject of their own extensive historical literature. There's no question that Pius made a number of public statements condemning racial and religious persecution and criticizing the Nazi regime for its treatment of the Jews. But there also seems to be little doubt that Pius could have acted far more forcefully than he did. Consider a single fact:

that this same Pius XII boldly excommunicated all members of Communist parties around the world in 1949 yet never took a similar step against the Nazis or other fascist movements.

The Vatican has staunchly defended Pius. Church historians have issued studies claiming that the pope acted aggressively to protest the persecution of the Jews and particularly to intervene of behalf on the Jews of Italy. They also point to many acts of courageous defiance of the Nazi regime by individual Catholic clergy and laity. However, most unbiased historians agree that these official documents, including "We Remember: A Reflection on the Shoah" (1998) and "Memory and Reconciliation: The Church and the Fault of the Past" (2000) are defensive, designed to exonerate Pius and the church through selective arguments that highlight the heroism of a few Christians while passing over the cowardice and complicity of many.

In any case, it would be simplistic to focus solely on the record of Pope Pius. The sad fact is that throughout its history, the Catholic Church prepared the way for Hitler and his exterminationist brand of anti-Semitism in a hundred ways great and small. By teaching that the Jews were Christ killers; by promulgating the notion of western Europe as Christendom, a political and social entity in which Christian beliefs and values must be dominant and in which all other forms of belief are by definition marginalized; and by permitting, even encouraging, priests and laypeople throughout Europe to spread anti-Jewish hatred and calls for violence, even from the pulpit—through these and a host of other sins the church did much to make Europe fertile ground for the harvest of blood that the Nazis ultimately reaped.

And recent scholarship continues to unearth evidence that the connection between Christian anti-Semitism and Nazism was even closer than this. For example, consider this passage from *The Popes Against the Jews* (2001) by David I. Kertzer of Brown University:

There is another uncomfortable truth that [the] official Church history of relations with the Jews obscures. The legislation enacted in the 1930s by the Nazis in their Nuremberg Laws and by the Italian Fascists with their racial laws—which stripped the Jews of their rights as citizens—was modeled on measures that the Church itself had enforced for as long as it was in a position to do so. Jews in the Papal States were still being prosecuted in the nineteenth century when caught without the required yellow badge on their clothes, mandated by Church councils for over six hundred years. As late as the 1850s, the Pope was busy trying to evict Jews from most of the towns in the lands he controlled, and forcing them to live in the few cities that had ghettoes to close them in. Jews were barred from holding public office or teaching Christian children or even having friendly relations with Christians. Church ideology held that any contact with Jews was polluting to the larger society, that Jews were perpetual foreigners, a perennial threat to Christians.

Against this historical backdrop, what moral legitimacy could the church have claimed in the battle against Nazism, even if its leaders had chosen to take a far more courageous stance than they did?

Even as they defend Pius and the other Catholics of his day, the scholars and archivists who control the Vatican's records of the period refuse to open the files to unbiased historians. This secrecy, which seems so strange more than half a century after the events in question, makes it difficult to fully judge the record of the church.

Nonetheless, Pius XII today even has his advocates for sainthood in the Vatican. His cause for canonization is being advanced by the Jesuit father Peter Gumpel, who is impatient with those who question Pius's legacy. I'm sad to report that Father Gumpel has even remarked that criticism of Pius "makes one wonder what the Jewish faction has against Catholics."

As a proud member of "the Jewish faction," I resent Father Gumpel's implication. But his comment is useful in one way. It highlights why the history of Catholic anti-Semitism, up to and including the time of the Holocaust, remains a living issue for Jews and Christians today. The attitudes and prejudices that produced the inquisitions, the expulsions, the pogroms, and the gas chambers have not disappeared—nor will they disappear on their own, without continuing spiritual and psychological vigilance by all peoples of goodwill.

Today, with Europe apparently on the verge of yet another spasm of anti-Semitic hatred like those that have convulsed the continent so often in its history, we need a Catholic Church that will stand up unambiguously against bigotry. It will take courage—greater courage, I'm afraid, than the leaders of the church have exhibited in the past.

BETWEEN TWO WORLDS

As it happens, I can speak to the Father Gumpels of the world from an unusual perspective. In a curious way, I can claim to be both a Jew and a Catholic. And in my own life, I have experienced much of the good and evil that are deeply intertwined in the connections between the two faiths. I know that it is possible for a Christian to perform acts of heroism and love in defense of Jews, even while harboring the same anti-Semitic attitudes that have motivated other Christians to murder.

How do I know these things? I know them because I was a "hidden child," one of thousands who survived the Holocaust through the intervention of a Christian guardian. It's a story with many remarkable twists and turns and one that illustrates, in microcosm, the fraught and tortuous nature of the Jewish-Catholic relationship.

My parents, Joseph and Helen Foxman, were married in Warsaw, Poland, in 1935. In 1939, with the advent of the Second World War, their lives were uprooted, like those of millions of other Europeans. When Warsaw was bombed, my parents fled eastward to Baranovich, in Poland (today located in Belarus), where my father's family lived. I was born there in 1940. Soon thereafter, in a further effort to stay ahead of the advancing German armies, our family moved east to Lithuania.

In the spring of 1941 the Germans occupied Lithuania, and an order went out for all Jews to report to the ghetto in Vilna, the nation's capital. My parents made a momentous decision. Rather than simply submit to the order, which they sensed might ultimately spell death for all three of us, they chose to leave me outside the ghetto, living with my nanny, a Polish woman named Bronislawa Kurpi. They assumed that this arrangement would be a short-term stratagem; nobody thought it would last four years.

Frau Kurpi was a humble servant girl, poorly educated and beset with physical infirmities, deaf in one ear and with a weak heart. But she was a devout Catholic, a tough-minded woman, and she loved me profoundly. She was willing to do whatever it took to keep me safe from the Nazis, even at the cost of risking her own life.

Like all Jewish boys—and unlike the Christian boys of that place and time—I was circumcised, a telltale sign of my religious heritage. So Frau Kurpi had to guard me zealously. I couldn't go out and play with the other kids; it wasn't safe for me to go to school. Frau Kurpi kept me by her side, and soon I was as devoted to her as if she had been my natural mother.

Naturally, she raised me as a Catholic. She had me baptized, she took me to church, she taught me the prayers, and she gave me a little cross to wear around my neck. I was a serious little boy, and I enjoyed sharing my nanny's faith. I've heard that whenever I passed a church, I would make the sign of cross myself, and when I met a priest on the street, I would stop to kiss his hand.

Frau Kurpi also raised me to despise Jews; after all, she shared the anti-Semitic assumptions of her world. During the years when my nanny took care of me, I would spit when a Jewish person walked by—"a dirty Jew," I would have said. As a small child of four or five, I couldn't comprehend the strangeness of this disconnection from my own heritage. There's a family story about a time I embarrassed Frau Kurpi on our way out of church by asking loudly, "Why is God barefoot? I guess he's poor. Maybe we should buy him a pair of shoes." She slapped me, and later she remarked, "It's your Jewish blood that makes you say such things."

During these years, while I lived with Frau Kurpi, my mother and father were never very far away. At first they both lived in the Vilna ghetto. From time to time they would visit me and my nanny. But I thought of Frau Kurpi as my mother (I called her *mamoushka,* "mommy"), and therefore I assumed that my mother and father were a sort of aunt and uncle. It was an unnatural, tense arrangement—and the more so because harboring a Jewish child outside the ghetto was a serious crime.

One evening during the summer of 1942 my father, who was then working in a chemical factory in Vilna, returned to his sleeping quarters in the city ghetto after seventy-two hours of back-breaking work, to find waiting for him a member of the ghetto's Jewish police.

The official informed my father that he had an order from his commander to bring him to police headquarters. There my father was confronted by a high official of the Lithuanian Gestapo, Court Examiner Julian Boyka. Also present were the prosecutor of the Jewish ghetto court and a ghetto policeman.

Placing a revolver on the table between himself and my father, Boyka warned him to tell the truth and then said, "You have and support, outside the ghetto, a boy of two or three years of age. The child stays at the home of a Polish woman by the name of

Bronislawa Kurpi. Don't try to deny it because I know everything about it."

Boyka then recounted Kurpi's activities on that very day, including a visit to my father at the factory, where my father had given her some soap. He then mentioned that a week earlier my father had given Kurpi some elastic to be used for garters. As he said these last words, Boyka placed his leg on the table. He was wearing garters made of the very material my father had given to the nanny.

Then Boyka dropped a bombshell. "Frau Kurpi is my sister. I had not seen her for more than ten years. I was certain that she was in Warsaw because that's where she was when the war broke out. But recently I met her, by accident, in the street here and was very happy to see her.

"She told me that she had gotten married and that she had a healthy, smart little boy. I took a liking to him as if he were my own child. So imagine my dismay when I discovered that this child, whom I love dearly and with whom I have so much fun, is a *Jewish* child."

Boyka went on to tell my father how much it hurt him that his sister could possibly have saved and raised a Jewish child, though "such a bright, good-looking child who understands everything and speaks like a grown-up." If the truth were discovered, it would cause him a lot of trouble, considering his position with the court. Therefore, he demanded of my father, "I want that child back in the ghetto within twenty-four hours. Otherwise, he'll be shot. I'll do it myself, even though I love the boy. And my sister will have to be shot for the crime of hiding a Jew."

Boyka added that my parents, as accomplices in the crime, would also have to be shot. But he would give them a chance to live. "Tomorrow evening at this time, I'll be here in the ghetto. I want to see that child right here in the police station. I'll also want two hundred gold rubles and a gold watch. Otherwise . . ."

All the time Boyka spoke, my father sat sphinxlike, trying not to betray his emotions. By the time Boyka asked, "What do you have to say?" my father had devised a plan.

He acknowledged that he knew Frau Kurpi, that she often came to visit him at the factory, and that her young charge was indeed a Jewish child. But, my father added, "Only one thing is not accurate: that the child is mine. He's not. He belongs to my wife's sister, who is also from Warsaw. Seven days before the war started, they were rounded up by the Bolsheviks and sent to Siberia. The child and his nanny—Frau Kurpi—happened to be outside Vilna that day, and so the boy was miraculously left behind."

My father added that he had asked the nanny to give the child to him and his wife but that she had refused, insisting that she would return the child only to his parents. Otherwise, she would raise him as her own. And the next time my father and Frau Kurpi met, he went on, "She took an envelope out of her purse and showed me a document stamped by the Catholic Church. It said that Bronislaw Kurpi had had her son baptized and named him Henryk Zeslaw Kurpi."

My father then expressed his willingness to retrieve young "Henryk" as Boyka had ordered, even though, he insisted, the child was not a Foxman. He said that he and my mother "will be ever grateful to you for getting back for us this member of our family who was taken away from us and baptized without the consent of his parents or anyone in his family."

However, as far as the two hundred gold rubles were concerned, my father told Boyka, "Not only do I not have them, but never in my life have I seen such a sum of money. And that is all I can tell you, sir."

Dismissed from the police station, my father spent the next day desperately seeking advice from others in the Jewish community. All agreed that I must be brought back into the ghetto and that

Boyka must be given all of my father's money, with whatever else the Jewish community could scrape together.

My parents then went to see the secretary of the ghetto police commandant, a lawyer named Abram Dimitrovsky, an honest decent man, who promised to help if he could.

Several hours later, when my parents returned to his office, Dimitrovsky told them that it would be best if Frau Kurpi and "Henryk" left Vilna immediately. Word was passed to them, and they escaped to a summer house five miles outside the city, where they went into hiding.

That evening, at the appointed hour, my father reported to the police station. Boyka was waiting. He had already been informed of Frau Kurpi's and "Henryk's" absence from the city, as well as of my father's dismal financial situation. He simply handed my father a document to sign, affirming that Joseph and Helen Foxman were childless and that they had no idea of the whereabouts of their "nephew."

My father's cool head—and Frau Kurpi's nerve—had made it possible for us all to escape with our lives, though narrowly.

Within a year, my mother escaped from the ghetto and established a false identity as an Aryan. She spoke Polish well and didn't look Jewish. She contacted my nanny and me and smuggled, stole, and worked to provide for us. My father was subsequently imprisoned in several concentration camps. He was one of the survivors freed when the Allied forces liberated Europe. Others in our family were not so lucky. Six members of my father's family and eight members of my mother's family were murdered by the Nazis.

In 1945 the Russians liberated Lithuania. These were happy times. Suddenly I was free to play and to go to school. I went to kindergarten, where I learned Russian, which now displaced Polish as the official language of education. I have just a few vague, happy memories of that childhood period. I remember the song we sang in class: "It's a blue ocean and a red train. . . . Come, let's travel to the farthest station." And I remember my teacher remarking, "When the sun is shining, Stalin is smiling; when it rains, he is angry."

There was very little housing in Vilna because of the destruction wrought by the war, and the few houses available were all controlled by the Soviets. So as a practical matter, all of us—my parents, Frau Kurpi, and I—moved in together and lived as a single family, even as a struggle began for control of my life and my future.

As far as I was concerned, I was a Catholic child. Perhaps I'd even begun to think about becoming a priest one day. I loved going to church with my "mommy," Frau Kurpi. But of course my parents were determined to reclaim me for my Jewish heritage—and for themselves, as my real family.

So my father began to carefully, gradually wean me away from Catholicism and back to Judaism.

He took me to synagogue for the first time in September of 1945, some four months after the liberation. It was the service of Simchas Torah, the joyous celebration of the Torah, filled with music, singing, and dancing. This was a wise choice on my father's part—he knew that a service like this would appeal to a small boy.

The story goes that, at the synagogue, a Soviet army officer who happened to be Jewish approached my father. "Is this a little Jewish child?" he asked. My father said yes.

His eyes damp, the officer replied, "In the past years, I have traveled thousands of kilometers across Europe without coming across a single living Jewish child. Please, may I dance with your son?" And so he lifted me up and danced around the floor with me, just as others were dancing around the floor with the scrolls of the Torah in their arms.

When we returned home I told my nanny, "I like the Jewish church, because they sing and dance there." Frau Kurpi, I am sure, was not happy. Her "son," she could sense, was beginning to drift away from her.

During the year that the four of us lived together, I attended both church and synagogue. Of course, I couldn't have explained the theological differences. I loved both sets of rituals. But little by little, my father began to substitute Jewish traditions for Catholic ones. For example, one day he gently took away the cross that hung around my neck and replaced it with a fringed prayer garment. I didn't mind the change; I imagine I thought that it was happening because I was growing up. The Latin prayers I'd learned from Frau Kurpi gave way to Hebrew prayers.

But Frau Kurpi didn't want to give me up. Rather than allow my parents to reclaim me for good, she decided that she would fight to keep me—both as her son and as a member of the Catholic Church.

Looking back on these strange events of almost sixty years ago, I think I understand some of the many motivations of my dear old nanny. Her Catholic piety certainly played a role. Frau Kurpi had a simple, sincere acceptance of the dogma that only a faithful Catholic could hope to be saved, and she wanted me to be part of the fold. But personal feelings of love and loneliness moved her as well. Since she had no husband and no children of her own, I think Frau Kurpi wanted desperately to keep me as her child. And so she began to take increasingly desperate steps to hold on to me.

First Frau Kurpi went to the local Soviet officials to accuse my father of having collaborated with the Nazis. This was a dangerous accusation, one that at worst could have led to his execution. He was arrested and interrogated, but there was no evidence for the false charge; after several days he was released.

Soon my father was appointed commissar (general manager) of a local factory. Frau Kurpi leveled a new accusation against him: stealing government property. Again he was arrested, and again he was freed after several days.

Then for a third time she denounced my father, claiming that he'd stolen soap and shoe polish from the factory. But by this time the authorities were no longer interested in her stories. One of the officials spoke to my father about it. "Mr. Foxman, this is getting ridiculous. We can't keep investigating you at the instigation of your nanny. We understand that she is trying to get you put away so that she can keep your child. You should really settle the matter once and for all."

"How can I do that?" asked my father.

"The only way is to take her to court and establish your parental rights legally."

So the result was a custody trial. What's most surprising is that Frau Kurpi, my parents, and I continued to live together even as this trial was unfolding.

Because I was only six years old, I had no voice in this trial. Had I been given one, I'm sure I would have chosen Frau Kurpi, whom I still considered to be my mother. But the court had to weigh the arguments of the three adults instead.

My parents' case, of course, was clear-cut. Little Abraham was their child, and he should be raised by them.

Frau Kurpi was forced to be inventive, and as a result, she used a variety of arguments. First she claimed that my parents were impostors—that they weren't really Joseph and Helen Foxman at all. This was easily disproved.

Then she claimed that I was really her illegitimate son and not related to the Foxman couple. Once again, my parents were able to produce witnesses that demolished the claim.

Finally, Frau Kurpi admitted that her first two statements had been lies. But she appealed for my custody on the ground that she was saving my soul for the Catholic Church. As you can imagine, the Soviets—official atheists—were not impressed by that argument.

In the end the court rendered the only logical verdict—that custody should be given to the natural parents. I still have the official document rendering that decision, which is written in Polish, Russian, and Lithuanian.

By now Frau Kurpi was bitterly angry, jealous, and resentful of my parents. But she wasn't quite through with her attempts to seize control over me. In a last-ditch effort, she had several of her relatives kidnap me and try to hide me in an apartment of theirs. Luckily, my parents were able to determine my whereabouts, and they got some Jewish acquaintances who lived in the same housing block to kidnap me back.

This was the last straw. My parents now realized that they had to get out of Europe once and for all.

The three of us packed our things and traveled to Austria. We made our way surreptitiously into the American zone and went to the displaced persons camp there, where we lived under American supervision for three years, waiting for visas to travel to the United States.

Remarkably enough, my parents stayed in touch with Frau Kurpi. They wrote to her from the camp and later from our new home in the United States, sending money and packages of goods. She never wrote back, and in 1958 the postal authorities informed us that she had passed away.

THE LOST JEWS

The story of Frau Kurpi is a strange one, in which compassion, love, faith, and heroism are inextricably blended with selfishness, deceit, cruelty, and pain. As I was growing up in my "new" Jewish family, I often wondered: Why is it that my nanny, who loved me so deeply, ended up hating my family? Years later, my father offered this explanation: "Frau Kurpi loved you, yes, but her love was excessive. And anything in excess is no good."

Despite all the trouble that Frau Kurpi caused for my family, I never heard a harsh word about her from my mother or my father.

For them, what mattered most was simply the fact that she had risked her own life to save me. Everything else, they forgave.

In 1992 the ADL helped sponsor a conference for hidden children in New York. About twelve hundred Holocaust survivors attended. Outside the meeting room, the halls were filled with billboards where faded, decades-old snapshots were posted, along with notes asking for any information about the people depicted. These were the plaintive cries for help of people who'd lost their childhoods.

I was luckier than most of the hidden children of Europe. My parents survived the Holocaust, and so they were able to fill in many of the gaps in my memories. But like other hidden children, I still have many questions about my past. Earlier in this chapter, I spoke about the unwillingness of the Vatican historians to open the Vatican archives. If the church's baptismal records were opened up, it would shed fascinating light on an important chapter in Jewish-Christian relations. It may be that the story told by these records would be an inspiring and heroic one. Tens of thousands of Jewish children may have been saved just as I was.

Still, the Vatican refuses to make these records public for reasons that have to do with politics and the instinctive desire of any hierarchical organization to protect and defend its own interests. After all, the church may "lose" thousands of Catholics if they release the records and reveal the Jewish heritage of many people who think of themselves as Christians.

As it is, we lose Jews every day—tens of thousands who may have been saved in Poland and other European countries but who have never known their own identity. We lose them because their foster parents who saved them are now in their eighties and nineties and are dying without telling them the truth—for religious reasons, to avoid exposing them to anti-Semitism, and perhaps out of pride, shame, or fear. It's one more sad legacy of the centuries of hatred separating two great faiths, a legacy I pray we can overcome in our lifetime, once and for all.

VICAR OF HOPE

In the last forty years some hopeful signs of change have emerged in the relationship between the Roman Catholic Church and the Jews.

The change began with the publication of the encyclical *Nostra Aetate* ("In Our Times"), promulgated by Pope John XXIII in 1965. One outgrowth of the heady reform era of Vatican II, *Nostra Aetate* represented a major break with the historic anti-Semitism of the Catholic Church. It officially absolved the Jews of the false accusation of having killed Christ, repudiated anti-Semitism, forswore persecution, and pledged a new commitment to interfaith dialogue.

For millions of Catholics and well-wishers around the world, Vatican II and the much-admired John XXIII represented a wonderful breakthrough for the church. The impact was especially strong in the United States, where it helped produce major strides in the effort by the American church to understand and respect non-Christian faiths.

However, Vatican II and the spirit of John XXIII did not have so profound an effect in the rest of the Catholic world. It remained for the current pope, John Paul II, to advance the cause of improved Jewish-Catholic relations in the spirit of *Nostra Aetate*.

I still remember my mother's reaction when John Paul II was selected in 1978 to be the first non-Italian pope in 455 years: "A Polish pope will either turn out to be the best pope in history or the worst." I'm happy to say that, at least in regard to Jewish-Catholic relations, he has been much closer to the best than to the worst. In fact, John Paul has done more for reconciliation between Jews and Catholics than anyone else in history.

Perhaps John Paul II is more sensitive than previous pontiffs to the issue of anti-Semitism because he grew up in Poland, the worst of the Nazi killing fields, during the war. He was born as Karol Wojtyla near Krakow in 1920 and studied religion secretly during the Nazi occupation. The years he later spent under the oppression of a rigid and authoritarian Communist state only intensified his commitment to freedom and his sympathy for victims of tyranny.

Those of us who know John Paul love him for his generosity, his gentleness, and his informal, open, accessible style. He is, literally and figuratively, a pope who sings. However, those within Catholicism who assumed that the pope's open manner suggested liberality in doctrine have been disappointed. He sees his role very clearly as keeper of the faith, with a strong emphasis on traditional disciplines and teachings. Some Catholics who would like to see their church change with the times consider him autocratic and severe. Paradoxically, however, this pope who is so strict with his fellow Catholics is more tolerant and open toward non-Catholics than any previous pope.

John Paul II has altered Jewish-Catholic relations in a host of ways. He has formally declared anti-Semitism a sin and was the first pope to do so. He has spoken of the Jews as the "elder brothers" of Christians, a sign of respect no previous pontiff offered. He has spoken of the Holocaust with greater empathy, respect, and a sense of personal responsibility and regret than we've heard from previous popes. He was the first pope to visit a synagogue (in Rome in April 1986) and later the city of Jerusalem. Most important of all, under his leadership, in September 1994, the Vatican finally recognized the state of Israel, whose existence is so essential for the continued vitality, normalcy, and even survival of the Jewish people.

I first met John Paul in 1979, just a year after his ascension to the papal seat. I participated in a joint mission to the Vatican regarding a Polish-Jewish-Catholic project—the establishment of what came to be called the Korczak Prize. Janusz Korczak (also known by the Hebrew name of Goldschmidt) was a much-loved Polish pediatrician and philosopher, sometimes affectionately referred to as the Doctor Spock of Poland. He wrote several books for parents, of which the most famous is probably *The Kingdom of Children.*

The Korczak Prize, created by the ADL in conjunction with the Polish American community, was given each year to honor a book that educates children about the evils of prejudice. We hoped to have the pope bless our endeavor, and we were delighted when our entourage was granted an audience with John Paul himself.

Like all first-time visitors to the Vatican, we'd been instructed in detail about how to behave in the pope's presence—how to stand, how to sit, what to say, and what not to say. The elaborate protocol led me to expect a man of great formality, perhaps a bit unapproachable. Instead, when we arrived, we found John Paul waiting at the door to greet us, which he did with the greatest imaginable warmth and courtesy. Here was a man who was very much at ease with himself and made everyone around him feel equally comfortable.

When I was introduced to the pope, I told him briefly about the story of my childhood rescue from the Holocaust by Frau Kurpi and how she raised me as a Catholic. "Holy Father," I said, "it would please me very much if you could ask for God's blessing on the soul of my poor nanny."

To my surprise, he embraced me and replied, "Thank you, my friend, for bearing witness to human compassion." And throughout the rest of our group's audience, he held on to my arm and wouldn't let go.

Co-opting the Holocaust

In the years since then, I've had almost every kind of audience with John Paul, in venues ranging from his summer retreat at

Castel Gandolfo to the rotunda in the Vatican as well as in Miami and New York; the pope has made it his practice to reach out to the Jewish community in each country or city that he visits. I've always used these meetings to raise important issues of Jewish-Catholic relations. This isn't easy to do. Papal visitors are required to tell the Vatican in advance what they want to talk about, and there is always some negotiation about what may or may not be discussed. Like most bureaucratic gatekeepers, the secretaries around the pope try to downplay sensitive items and avoid potential controversies. However, most of the time I've been able to use my audiences with the pope to make the points I felt compelled to make without unduly straining our relationship or stepping across the lines drawn by etiquette and diplomacy.

Only once in my relationship with John Paul were we unable to come to an agreement concerning what might be discussed in an audience. It was in 1989, during the controversy over the establishment of a Carmelite convent at Auschwitz. The story vividly encapsulates the lingering hurts in the Jewish-Catholic relationship, showing how even the well-intentioned gestures of the church sometimes convey an attitude of arrogance and incomprehension toward Jewish sensibilities that makes it very difficult for us to feel reconciled with our Christian neighbors.

From a Catholic point of view, there were many reasons a Christian center of prayer ought to be created at Auschwitz. For one thing, it would serve as a symbol of Christian repentance for complicity in the Nazi crimes. For another, it would commemorate the thousands of Christians who also were murdered at Auschwitz and the other extermination camps. Thus there was a certain logic to the idea of creating a convent at this historic site.

However, we Jews were deeply troubled by the existence of the convent, and especially by the erection of a large cross nearby that practically overshadowed the entrance to the camp. We feel this way not because we are offended by Christianity or its symbols, but because we consider this symbolic gesture part of an ongoing effort by the church to *universalize* the tragedy of the Holocaust.

What do I mean by universalizing? As most people know, the Nazis vented their hatred not exclusively on the Jews but on other disfavored groups, including gypsies, Communists, homosexuals, and the mentally disabled. Universalization seizes on this fact to depict the Holocaust not as a uniquely Jewish tragedy but as a generically human one.

The impulse behind universalization isn't entirely wrong. If all human beings are brothers and sisters, then there is a sense in which all humans share in the suffering of any group. But the way in which the church, or some leaders within the church, have used the concept of universalization carries dangerous and false implications. If *everybody* was a victim of the Nazi violence, then no one group can be singled out for blame. And specifically, the perpetrators of the Holocaust cannot be identified as Christians. Thus universalization has the subtle effect of exonerating the church and its members for their role in mass murder.

Admittedly, the Nazis and their helpers who committed the murders did so not as Christians but rather as functionaries or supporters of a political regime. Yet it's also true that many Germans and Austrians who spent the week murdering Jews then went to church on Sunday, apparently seeing no inconsistency in their actions. And it's also true that the political and social atmosphere in which the persecution and killing of millions of Jews could be seen as broadly acceptable could not have existed without the tacit acceptance of the Christian churches—as well as the ingrained anti-Semitism of twenty centuries of dogma, doctrine, and preaching that demonized Jews.

As Elie Wiesel puts it, not all the victims of the Holocaust were Jews, but all Jews were victims. And because Christians and the Christian churches had spread hatred of Jews for so long, it's impossible for Christians—and in particular the Catholic Church—to regard themselves as passive or innocent bystanders during the Holocaust. The killings could not have happened without the sins of millions of Christians—sins of commission as well as sins of omission.

Furthermore, by seeking to universalize the Holocaust within a Christian theological framework, likening the murder of the Jews to the sacrificial death of Jesus and using the cross as the over-arching symbol for these acts and their meaning, the church is unwittingly denigrating and denying the Jewishness of the sufferers at Auschwitz. In effect, it is making them into honorary Christian martyrs—as if forcibly converting them after death.

With the world debating these issues in relation to the convent at Auschwitz, I felt it was important for me to offer the pope a Jewish perspective on the issue—to explain, from the heart, why many Jews found it disrespectful to the memory of those killed at Auschwitz to commemorate their suffering with the Christian symbol of the cross. But the Vatican officials forbade me even to mention the issue. Under the circumstances, I felt compelled to cancel our audience, making it clear why I was doing so.

I later learned that there was a special reason for the pope's reluctance to discuss the Auschwitz issue at that time. As the world now knows, John Paul was involved behind the scenes with the Solidarity labor movement and the anti-Communist effort in Poland. From his contacts behind the Iron Curtain, he understood something that few people did—that the Soviet empire was on the brink of collapse, with popular movements throughout Eastern Europe ready to step forward and take power. At such a sensitive moment in history, the pope was striving to avoid public comment on *any* issues related to Poland. Thus his silence over Auschwitz.

Eventually, the pope did intervene in the convent controversy. He worked out a compromise under which the nuns moved to a new building in 1994. However, the central symbol of the controversy—the large cross, which had actually been used during a papal mass in nearby Krakow in 1979—remains in place at Auschwitz, to the dismay of many Jews. For many Poles, it is a symbol not only of Christian faith but of their national heritage and their pride in being the homeland of John Paul II. But to the Jews of the world, it is a painful and unnecessary reminder of past hurts and present insensitivity.

There have been other times during the papacy of John Paul when I felt compelled to criticize him. When he visited Damascus in Syria in May 2001 he was greeted by President Bashar Assad, whose public remarks at the time included a number of anti-Semitic statements, including a reference to Jewish "persecution" of Christ. To my disappointment, the pope was silent about these comments.

This is precisely the kind of acceptance of hatred that the ADL must rebuke. We took out a newspaper ad faulting the pope's silence. Many leaders of the Catholic Church responded with anger.

For a time our dialogue with the church was broken off, which I regretted. But the rift was soon healed. When the twentieth anniversary of the papacy of John Paul was celebrated in 1998, we took out ads praising him for his efforts on behalf of world peace, brotherhood, and tolerance. And when he visited Israel in March 2000, we published ads in Hebrew and English, welcoming him to Israel and spelling out the good he had done for Jewish-Catholic relations. Thus some balance was restored to our relationship, and soon we were on speaking terms again.

RUSHING TO SANCTIFY

Has John Paul II done all that could or should be done to redress the church's past sins against the Jews? Probably not. Certainly some observers find fault with his efforts. After the pope's visit to Israel in 2000, Rabbi Arthur Hertzberg criticized him and groups like the ADL (as well as other Jewish groups) that praised his outreach efforts.

On the whole, John Paul II has been an important force for good where Jewish-Catholic relations are concerned. But the tragic legacy of two thousand years of hatred can't be erased within a generation. There are still many unfortunate irritants in the Catholic-Jewish relationship, some of which revolve around the church's process of beatification and sainthood.

In 1998 the church canonized (that is, elevated to sainthood) a Holocaust victim named Edith Stein. Born a Jew, Stein converted to Christianity, became a Carmelite nun known as Sister Benedicta of the Cross, and then died in a concentration camp. Her story is a tragic one, of course. But it's far from clear that she was really a *Christian* martyr. Edith Stein was killed not because of her Catholic beliefs but because of her Jewish heritage. Many Jews and others see her canonization as another subtle attempt by the church to redefine the Holocaust within a Christian context and thereby de-Judaize it.

Stein's is not the only controversial name on the roster of new Catholic saints. Another is that of Father Maximilian Kolbe, a priest who operated a number of newspapers and magazines on behalf of a Catholic mission center in Poland. Like Stein, Kolbe was killed in the Holocaust, and in a fashion that was undoubtedly heroic: when the SS officers in charge of Auschwitz decided to execute every tenth prisoner on Kolbe's block as punishment for an escape, Kolbe volunteered to take the place of a worker with a wife and children. Kolbe was canonized in October 1982. Why are Jews uncomfortable with this? Because of the anti-Semitic views contained in the papers Kolbe supervised and even in some of his own writings.

Today there are several controversial candidates for sainthood. As noted earlier, one of these is the wartime pontiff, Pope Pius XII.

Another is Cardinal Alojzije Stepinac of Croatia, who supported the pro-Nazi Ustashi regime during World War II (although he privately protested the mistreatment of the Jews). Stepinac was beatified in 1998, the final intermediate step before sainthood.

I would ask the pope to postpone canonizations like these, which will cause pain to the Jewish community, and in particular to Holocaust survivors and their families. After all, there is no reason to rush; the responsibility of the church is to make decisions that will stand for an eternity. I understand the importance of the canonization process for Catholics; it's a vehicle by which the church connects religion more intimately to the people. But the four or five troublesome canonizations on John Paul's agenda can and should wait for a more opportune time.

WHITHER THE CHURCH?

Sometimes I worry about how the worldwide Catholic Church will change after John Paul's inevitable passing from the scene. Much depends, of course, on the beliefs and aspirations of the person chosen as his successor. However, I suspect that in the years to come the focus of the Vatican on Jewish relations will diminish. The lapse of time, and the inevitable passing of those who remember the horrific events of the 1940s, may mean that these issues will fade from public awareness without having been dealt with fully, honestly, and completely. Already, I imagine, some within the Catholic hierarchy must feel, "We've given the Jews enough attention for a while."

In a way, this is natural. Like most large, bureaucratic organizations, the church responds to the issues that are thrust in its lap. Over the last twenty years issues related to the Jews got a great deal of attention. In the near future, they are likely to get lower priority than other urgent matters. I can imagine that the

Catholic relationship to the burgeoning challenge of the Muslim world may take center stage for a while.

Future progress in Jewish-Catholic relations will depend on the various national churches, which have a lot of leeway as to what they emphasize and what specific practices and policies they follow. The ADL will be following these developments closely, especially in Latin and Central America, where there is still much to be done to free the church from its legacy of anti-Semitism.

In North America the spirit of Vatican II is widespread, including an attitude of relative openness toward other faiths. However, in Latin America, Africa, and Europe, John Paul's message of understanding and tolerance has yet to be fully accepted. Thus the Catholic Church in these regions remains more open to anti-Semitism than the American church. Every Easter there are still reports of anti-Semitic sermons being preached from Catholic pulpits around the world (and sometimes even in the United States). There appears to be an especially high level of anti-Semitism in Spain and the Hispanic countries generally. There's no question that traditional religious teaching—specifically, the discredited charge of deicide—plays a role.

Even more disturbing is the role played by some avowedly Catholic politicians and parties in promulgating intolerant, far-right doctrines:

- In Austria, extreme right-wing groups like the Austrian Freedom Party (Freiheitliche Partei Österreichs, or FPO) purvey a blend of Catholic fundamentalism and anti-Semitism, blaming conspiracies led by Israeli and American Jews for the problems of modern society.

- In France, the Catholic fundamentalist groups Chrétienté-Solidarité and Jeunesse Action Chrétienté advocate denial of citizenship to non-Christians and espouse an extreme pro-Palestinian position that, in effect, calls for the destruction of the state of Israel.

· In Poland, despite official pronouncements about the abandonment of traditional anti-Semitism, Radio Maryja, directed by the popular father Tadeusz Rydzyk, broadcasts a diet of strongly xenophobic propaganda under a Catholic veil, denouncing "Jews, global capitalism, representatives of other religions [beside Catholicism], and secularization" for an audience of over five million.

These and other bigoted groups that claim the mantle of Christianity offer an ideal opportunity for church leaders to exhibit the moral courage that many of their predecessors failed to show. The world is watching and waiting.

The political, social, and psychological impulses that drive today's conservative movements around the world are not inherently allied with intolerance or bigotry. At its best, conservatism stands for the preservation of institutions and customs that help society strike a reasonable balance between order and liberty. It also stands for the defense of traditional moral values that most cultures have embraced. These are goals that almost everyone can respect, even while we debate the proper means for achieving them.

Unfortunately, the conservative movement has too often been co-opted by those who would use "traditional values" as a veil for racial, religious, and ethnic hatreds. The right-wing European organizations I cited above are examples of this perversion of the honorable conservative tradition. In the next chapter we'll look at the same phenomenon in America and how supposedly conservative organizations are actively fomenting the spirit of anti-Semitism in the United States.

CHAPTER 4

Danger on the Right:
Violence and Extremism in the American Heartland

FOR MOST AMERICANS September 11, 2001, has served as a psychological watershed, a before-and-after date marking a dramatic change in our national consciousness. As a result, another traumatic event—the bombing of the Alfred P. Murrah Federal Office Building in Oklahoma City, which killed 168 people in April 1995—has been cast into historical shadow. But we forget that horrific crime and its lessons at our peril.

Oklahoma City brought into view a landscape of heavily armed, fanatically antigovernment superpatriots, white supremacists, and neo-Nazis—disaffected loners crisscrossing the gun show circuit and camouflaged paramilitaries training secretly in the woods. These new militants, angrier and more volatile than the extremists of the past, are bent on attacking America in order to save it, no matter how great the "collateral damage."

Unfortunately, in the years since the bombing, the media focus on these far-right extremists has wavered. At first Timothy McVeigh, the man convicted of the crime and executed on June 11, 2001, drew the lion's share of press attention. McVeigh left an unnerving impression in part because he did not fit the usual stereotype of a right-wing extremist: the coarse, bellicose, race-hating workingman. McVeigh's family was middle class. He was articulate and polite, never ranting, and not overtly driven by bigotry. To observers of the right wing, even his belief in the evils of the federal government was unexceptional. McVeigh's anonymously American quality made him seem incommensurate with the pain and damage he caused. Even his biographies implied that McVeigh represented, in some fashion, a ghastly underside of the American dream: titles included *American Terrorist, All-American Monster,* and *One of Ours.*

Ultimately, McVeigh's most notable personal quality—his terse, military, inhuman single-mindedness—though repellent, lulled many Americans into a false sense of security. For McVeigh appeared to be unique in the intensity of his convictions and his willingness to sacrifice himself. His acknowledged guilt and his claim to virtually total responsibility for the crime helped reinforce the assumption that life in America is essentially free from terror. McVeigh himself became so identified with Oklahoma City that his few confederates either faded into oblivion (as with Michael Fortier, with whom he discussed his plans and who testified against him) or came to be seen as merely nominal (as with coconspirator Terry Nichols).

Meanwhile, other tragic outbreaks of extremist violence in America continued throughout the late 1990s, receiving only sporadic media coverage. Antiabortion radicals committed several murders and many bombings and harassments, intimidating women seeking abortions and those who work at clinics. Militia groups engaged in shoot-outs with federal agents, and hate crimes

targeting gays, Blacks, Jews, Asians, and other minorities continued. But most Americans dismissed the extremists who committed these crimes as a minor irritant rather than a threat to
national security.

September 11 shattered this complacency. The mass deaths and
massive destruction of the attacks on New York and Washington,
together with the possibility of even greater dangers, brought
extremist terror into the forefront of Americans' consciousness.
Within a day it had forced itself into our routines and into the
way we view our nation and the world. More thoroughly than at
any time since the Second World War, we have been forced in
this country to consider the constitutional liberties we have
inherited and to work consciously to safeguard them.

It is too facile to label the events of September 11 as merely an
attack on American democratic values, but there is no question
that America's exceptionally free society attracts all manner of
hatreds. And although attention has been focused on Islamic militants, homegrown extremist movements in America have also
been galvanized in the wake of September 11. Many fringe groups,
even those usually associated with hatred of foreigners and non-
Christians, were seemingly inspired by the powerful display of
organization and sheer will to destruction exhibited by the
extremist Islamic terrorists.

One American neo-Nazi said of the September 11 terrorists,
"We may not want them marrying our daughters . . . but anyone
willing to drive a plane into a building to kill Jews is all right by
me. I wish our members had half as much testicular fortitude."
Others on the extreme American right share these sentiments;
some have acted or threaten to act on them; and all join with foreign terrorists in despising the ethos of tolerance, pluralism, and
the rule of law that defines American life.

That's one reason the danger posed by these homegrown terror
groups has never been more serious. Today, when the attention of

law enforcement organizations and ordinary citizens is distracted by dangers from abroad, we may again be vulnerable to a threat from the American heartland.

At ADL we've been working to monitor, warn against, and limit the influence of hate groups for a long time. In fact, in the fall of 1994, some six months prior to the Oklahoma City bombing, ADL published the first major warning about the dangers of extreme right-wing terrorism in the United States. Titled *Armed and Dangerous,* our report described the emerging antigovernment militia movement, discussed its espousal of anti-Semitism, racism, and other forms of bigotry, and documented the likelihood of violent attacks against government and other targets in this country.

The report received little response in the mainstream media. Few Americans were aware of the growing power of the militia movement—and even fewer, it seemed, wanted to know. When the horrific news about Oklahoma City broke, many commentators quickly blamed Arab terrorists. ADL's response was, "Let's not rush to judgment." Sure enough, within twenty-four hours it had become apparent that the bombing was the work not of Islamic fundamentalists but of American ultranationalists like those profiled in *Armed and Dangerous.*

When the name of Timothy McVeigh surfaced as the lead suspect, we at ADL were among the few people who recognized it. As part of our ongoing mission to track the doings of hate groups, we'd been monitoring an anti-Semitic tabloid newspaper called the *Spotlight,* published by the extremist organization Liberty Lobby. Using a pseudonym, McVeigh had run an ad in the *Spotlight* offering bazookas, mace, and other weapons for sale. Tracking leads in the immediate aftermath of the bombing, we traced the ad to a post office box in Kingman, Arizona, where McVeigh was living in a trailer. This connection helped lead to McVeigh's capture by law enforcement authorities. And when the media realized that the ADL had been on the story before most reporters, *Armed and Dangerous* suddenly became required reading in newsrooms around the country.

In the years since Oklahoma City, ADL has continued to monitor America's right-wing extremists. We've also continued to work with law enforcement officials and civic organizations to thwart the extremists' dreams of violence—for example, by drafting a model statute that restricts paramilitary activities by unofficial groups, which has become law in Texas and several other states.

In the next few pages I'll sketch what we know about the threats posed by the current crop of extremist organizations in America.

AN EVER-SHIFTING NETWORK

The world of right-wing extremism is a complex, ever-shifting network of groups, individuals, and movements. It includes traditional hate groups such as the Ku Klux Klan and various neo-Nazi organizations; quasi-military groups often described as militias; splinter churches that espouse perverted forms of Christian doctrine; and an array of political and social organizations that promote hate through publications, conventions, and meetings.

It also includes "common law court" groups, which seek to replace our legal system with one of vigilante justice. These pseudojurists render unenforceable judgments regarding genuine legal disputes and issue phony legal documents, including property liens and criminal indictments, in order to intimidate and defraud their enemies. The Republic of Texas, a San Antonio–based group that claims to represent the true government of the state, actually combines the militia and common law court movements in a single frightening organization.

Finally, right-wing extremism also includes elements of such unorganized movements as the skinheads, young people who express individual alienation and anger through styles of dress and music. Some skinheads have espoused anti-Semitism and racism, and a few advocate violence.

Traditional racist and anti-Semitic ideas are given continued currency by the activities of far-right extremist organizations. For example, among the classics of anti-Semitic literature is *The Protocols of the Learned Elders of Zion,* which supposedly details the secret Jewish plan for world domination. Scholars have long known that the *Protocols* was forged nearly a century ago by the secret police of czarist Russia. Nonetheless, the book has been reprinted dozens of times in many languages around the world, and it is currently hawked at the many common law court and militia-oriented events held around the United States every year.

Other propaganda materials distributed at meetings, conventions, gun shows, and over the Internet denounce Jews and Jewish institutions as coconspirators in the establishment of a tyrannical "new world order." Believers in this theory see the federal government working in league with the United Nations to strip Americans of their rights, using gun control laws to render patriotic citizens helpless to defend themselves. Because they view law enforcement agents as the foot soldiers in a federal plot to impose tyranny, they refuse to recognize the authority of law enforcement, leading to heated, and occasionally deadly, encounters with local, state, and federal authorities.

On the outermost fringe of the far-right extremist movement are independents who operate beyond the purview of organized militia groups. They work in tiny, loosely organized cells to achieve their ultimate goal—the overthrow of the United States government. Many are fervent believers in the tenets of Christian Identity, a racist religious creed that advocates violence in defense of white power. For example, the members of the so-called

Phineas Priesthood support a violent credo of vengeance advocated by Identity leader Richard Kelly Hoskins in his book *Vigilantes of Christendom: The Story of the Phineas Priesthood,* which perverts passages of the Bible, including the story of the zealot Phineas, to justify racist and anti-Semitic violence.

After losing popularity in the late 1990s, Phineas Priesthood activity once more became common among right-wing extremists in early 2002. Some of the propaganda offers chilling evidence of a new rapprochement between homegrown American extremism and Islamic terrorism. In April 2002, for example, Joshua Caleb Sutter, leader of the Pennsylvania branch of the Aryan Nations hate group, praised Palestinian suicide bombers who killed Israeli civilians, predicted America would also soon see such bombers, and expressed hope that "Phineas Priests and Priestesses" would soon awake to join Islamic extremists in executing vengeance. Soon the Aryan Nations Web site began openly recruiting people to become Phineas Priests with a page that prominently displayed images of Osama bin Laden and a suicide bomber.

Hoskins is only one of the writers whose work has influenced the recent development of far-right extremism. Perhaps the most famous is William Pierce, whose *Turner Diaries,* self-published in 1978 under the pseudonym of Andrew Macdonald, helped inspire Timothy McVeigh. (A copy of the *Diaries* was found in McVeigh's trailer when he was apprehended after the Oklahoma City bombing.) The book depicts the violent overthrow of the federal government and, in a scene that chillingly prefigures Oklahoma City, the preparation of a bomb designed to destroy the national headquarters of the FBI. Also depicted is the systematic killing of Jews and nonwhites in order to establish an "Aryan" world, which Pierce justifies in this way:

If the White nations of the world had not allowed themselves to become subject to the Jew, to Jewish ideas, to Jewish spirit, this war would not be necessary. We can hardly consider ourselves blameless. We can hardly say we had no choice, no chance to avoid the Jew's snare. We can hardly say we were not warned.

Inspired by such writings, other McVeighs are continually looking for opportunities to commit similar crimes inspired by hatred. Some have targeted Jews, while others have targeted Blacks, Asians, gays, abortion providers, and other "enemies."

SUMMER OF HATE

Violent assaults by right-wing extremists tend to come and go in waves. Perhaps the worst onslaught of such attacks in recent years occurred in 1999, during a horrific period often called "The Summer of Hate." Three major episodes of violence dominated the headlines that summer.

In mid-June, three synagogues and an abortion clinic in Sacramento, California, were firebombed. The crime was eventually traced to Benjamin Matthew Williams and James Tyler Williams, brothers connected with the racist World Church of the Creator. A month later (July 21), the Williamses shot and killed a gay couple. Both brothers were arrested and sentenced to long jail terms; Benjamin committed suicide in prison in November 2002.

On the Fourth of July holiday weekend, Benjamin Smith, another member of the World Church of the Creator, went on a Midwestern shooting spree. It started with an attack on Jews departing a synagogue in a suburb of Chicago, Illinois. Then Smith shot and killed the well-known Black basketball coach Ricky Birdsong, and finally he killed a Korean American graduate student named Won-Joon Yoon in Indiana. Smith was identified by local authorities with the help of ADL's Chicago office, which had long monitored his activities; a police chase climaxed in Smith's suicide.

In August, Buford Furrow Jr., who had worked as a security guard at the annual gatherings of the racist Aryan Nations organization, shot and killed Joseph Ileto, a Filipino American postal worker in Los Angeles. Furrow then entered a Jewish community center and began spraying bullets around, wounding several people, including a small child. Furrow, who declared that his action was intended as "a wake-up call to white America to start killing Jews," is now serving a life sentence in federal prison.

For many Americans the Summer of Hate was a reminder of the deadly impact that bigoted speech and writings can have. Each of the assaults was directly linked to specific extremist organizations and individuals that preach hatred and violence against a wide range of minority groups. The perpetrators obviously took these messages of hatred very seriously, launching assaults indiscriminately against Jews, Blacks, Asians, gays—anyone and everyone who could be considered different and therefore fit for extermination.

This pattern of attacks underscores the importance of ADL's long-standing commitment to fight bigotry of all kinds, no matter who the target may be. Yes, ADL has a special mandate to focus on anti-Semitism and violence directed against Jews. But we are equally vocal about hate speech and hate crimes against all other racial, ethnic, and religious groups as well as hatred based on gender or sexual orientation. Experience shows that bigots who hate one group of people are overwhelmingly likely to extend the same hatred to others. Thus no one can afford to ignore bias attacks on other groups in society, because if someone else can be assaulted with impunity, the chances are excellent that you or I may be targeted next.

Unfortunately, we don't know when or where the next attack will occur. We only know that it *will* happen, because incitements to hatred and violence have never been more widespread than today.

There is a wide range of organizations and individuals currently spreading racist and anti-Semitic hatred throughout the United States. In the pages that follow I'll tell the stories of three of the most prominent groups: Aryan Nations, the National Alliance, and the Council of Conservative Citizens. They illustrate both the variety of styles on America's far-right fringe and the shifting connections and alliances that link these dangerous extremist groups—and even give them access to the supposedly respectable mainstream of American opinion and politics.

APPEALING TO WHITE PRIDE: THE ARYAN NATIONS MOVEMENT

Aryan Nations is one of the country's best-known enclaves of white nationalism, a movement that asserts the need for white Christians to retake control of the United States, using violence against Blacks, Jews, and other minority groups if necessary. Founded as an outpost of the Christian Identity religious movement, the organization also incorporates neo-Nazi themes; its founder and longtime leader, Richard Girnt Butler, openly adulates Hitler. It is no surprise, then, that Aryan Nations for many years has shared members with several other white supremacist and neo-Nazi groups. Furthermore, the Aryan Nations compound at Hayden Lake, Idaho, has served as one of the central meeting points and rallying grounds of far-right extremists of all stripes.

Born in 1918, Butler is a World War II veteran who later worked as an engineer for Lockheed in southern California, where he was introduced to Christian Identity teachings by William Potter Gale, a retired colonel and onetime aide to General Douglas MacArthur in the South Pacific. By the mid-1960s, Butler had fully embraced Identity and served as national director of the Christian Defense League, an organization founded by the

most prominent popularizer of Identity, Wesley Swift. Butler
worked under Swift for ten years until Swift's death in 1971, at
which time Butler proclaimed his Church of Jesus Christ
Christian to be the direct successor to Swift's ministry.

Butler moved the congregation to northern Idaho where it
became, in his words, a "Call to the Nations" or Aryan Nations.
Its goal, as a newsletter stated, was to form

> a national racial state. We shall have it at whatever price is neces-
> sary. Just as our forefathers purchased their freedom in blood so
> must we. . . . We will have to kill the bastards.

Several Aryan Nations associates have acted on this call to
arms. During the early 1980s, for example, Butler followers
joined with members of the neo-Nazi National Alliance and Ku
Klux Klan splinter groups to form the Silent Brotherhood,
known more widely as the Order, which planned to overthrow
the United States government in hopes of establishing an Aryan
homeland in the Pacific Northwest. In order to raise funds for
this revolution, members of the group went on a crime spree in
1983–1984 that included bank robberies, counterfeiting, bombings, armored car holdups, and murder. The counterfeiting operation was based at the Aryan Nations compound.

Ostensibly, the Order's activities came to an end in December
1984 when its founder and leader, Robert J. "Bob" Mathews, died
in a fire during a shoot-out with federal agents on Whidbey
Island, Washington, and many of its members were caught and
incarcerated. Yet the Order, and to a lesser degree Aryan Nations,
has retained a mythic status in the far-right underground. Its leg-
end is now perpetuated through the Internet, inspiring a new
generation of would-be white revolutionaries and further rein-
forcing the Aryan Nations brand.

On its Web site, Aryan Nations blends anti-Jewish sentiments with opposition to the American government in an Aryan Declaration of Independence, which cruelly parodies the original:

> The history of the present Zionist Occupied Government of the United States of America is a history of repeated injuries and usurpations, all having a direct object—the establishment of an absolute tyranny over these states, moreover, throughout the entire world. . . . We, therefore, the representatives of the Aryan people, in council, appealing to the supreme God of our folk for the rectitude of our intentions . . . solemnly publish and declare that the Aryan people in America are, and of rights ought to be, a free and independent nation. They are absolved from all allegiance to the United States of America, and that all political connection between them and the federal government thereof, is and ought to be, totally dissolved; and that as a free and independent nation they have full power to levy war, conclude peace, contract alliances, establish commerce, and to perform all other acts which independent nations may of right do.

The declaration concludes by quoting the so-called Fourteen Words, coined by David Lane of the Order: "We must secure the existence of our people and a future for white children." This sentence has become a popular battle cry for white supremacists and neo-Nazis around the country.

During the 1970s and 1980s, Butler raised funds for Aryan Nations activities by encouraging congregants to make offerings and pay membership fees, in addition to selling flags and tapes of his sermons. Supporters are also required to tithe 10 percent of their incomes. The group's financial prospects changed dramatically in 1998, however, when Carl E. Story and R. Vincent Bertollini, acquaintances of Butler who had become wealthy in the field of computer technology, donated a significant sum of money to the group. Both recent transplants from California's

Silicon Valley to Sandpoint, Idaho, they founded and now lead the Eleventh-Hour Remnant Messenger, a Christian Identity ministry that shares the apocalyptic racism of Aryan Nations. The two men have underwritten several expensive propaganda efforts, including the distribution of a videotaped interview with Butler that was reportedly sent to nine thousand households in northern Idaho.

In September 2000, Aryan Nations made national headlines again when a jury awarded $6.3 million to Victoria and Jason Keenan, a mother and son represented by the Southern Poverty Law Center who had been assaulted, chased, and shot at by Aryan Nations guards after briefly stopping their car on a road in front of the group's compound two years before. The jury found Aryan Nations and Butler guilty of negligence in the selection, training, and supervision of their security guards. The judgment bankrupted Butler and his group, and the twenty-acre Aryan Nations compound and the Aryan Nations name were legally handed over to the Keenans.

Butler renamed his organization the Aryan National Alliance. Patrons Bertollini and Story purchased a new home for the Identity pastor in nearby Hayden, Idaho, where he vowed to continue his activities, including propaganda distribution and posting to the Web site.

In September 2001 Butler announced that he had chosen Ray Redfeairn, a former Ku Klux Klan member from Ohio, as his successor. The selection suggests that the group will remain militant and possibly volatile: Redfeairn has a substantial record of criminal activity, beginning long before his 1997 conviction on weapons charges. In 1979 he shot a police officer several times during a traffic stop, then pleaded guilty by reason of insanity (psychologists described him as a paranoid schizophrenic during the trial) and spent four years in a mental hospital. In 1985 he

pleaded guilty to attempted aggravated murder for the shooting
and to charges of aggravated robbery for an incident that
occurred prior to the shooting. He was released from prison in
1991.

A number of Aryan Nations members have recently left the
group, some to form their own organizations. For example, two
former Aryan Nations supporters have announced the formation
of a new group called Church of the Sons of Yahweh, headquar-
tered in Ohio. Disturbingly, both Aryan Nations and the new
Church of the Sons of Yahweh have adopted the Phineas
Priesthood's concept of a violent class of white "warriors for
God." Thus factional infighting and competition for members
may lead to an increased risk of violence or even terrorism by past
and present members of Aryan Nations.

REACHING OUT TO YOUTH: THE NATIONAL ALLIANCE

Historically, the National Alliance has its roots in the Youth for
Wallace campaign, established by Willis Carto, the anti-Semitic
founder of Liberty Lobby, in support of the 1968 presidential bid
of Alabama governor George Wallace. After Wallace's defeat,
Carto renamed his organization the National Youth Alliance and
tried to recruit college students to his increasingly radical cause.

In 1970 William Pierce, later renowned as the author of *The
Turner Diaries,* joined the National Youth Alliance. Pierce had been
an associate of George Lincoln Rockwell and the American Nazi
Party (ANP), and with the help of several former ANP activists
he ultimately led the organization away from Carto's influence.

By 1974 the organization had split into separate factions, and
Pierce's wing became known as the National Alliance (NA).
Until his death in July 2002 Pierce ran the group and edited its
magazine, *National Vanguard* (originally titled *Attack!*), as well as an

internal newsletter, *National Alliance Bulletin* (formerly called *Action*). He was also in charge of the group's "American Dissident Voices" weekly radio address, and he controlled other businesses associated with the NA, including National Vanguard Books, Resistance Records, and Cymophane Records.

In the early years of NA, Pierce held weekly meetings near Washington, D.C., in an effort to attract people to the organization. In 1985 Pierce relocated to a 346-acre farm in Mill Point, West Virginia, which he bought for $95,000 in cash; he called his new compound the Cosmotheist Community Church. There has been some speculation over the years that at least some of the money used for the purchase came from the proceeds of bank and armored-car robberies committed by the Order, the white supremacist terrorist gang described earlier.

As of December 2001 the NA had more than thirty-five cells from coast to coast, and there has been evidence of NA activity in about thirty states across the country. While other extremist hate groups appeal to a narrower range of followers, the NA's membership varies widely in terms of class and age. Some of the group's followers are young racist skinheads, while others are middle-aged, upper-middle-class men or couples. Moreover, Pierce boasted that several judges and professors were members of the organization.

National Alliance leaders are known for their energetic recruiting and are continually in search of innovative ways to advance their white supremacist message. To attract new followers, NA leaders and members have used billboards, hung organizational banners in prominent locations, rented booths at gun shows, posted their propaganda materials on public property, and distributed NA literature in suburban neighborhoods and on college campuses. One popular item distributed by the NA on high school and college campuses has been *The Saga of White Will!!,* a racist, anti-

Semitic comic book that encourages students to join the fight for "nationalism and racial and ethnic self-determination everywhere." In North Carolina an NA member even runs a stock car bearing the NA Web site address during weekly auto races.

William Pierce tightly controlled the NA's message, requiring followers to obtain his permission before they spoke publicly or created new propaganda materials. Pierce's strict enforcement of these rules helped the National Alliance conduct its activities with little of the internal politics and strife that have sapped the strength of other hate organizations. Under his guidance it developed one of the strongest brands in the extremist world.

Essential to the group's vision of the future is the creation of White Living Space, an area that incorporates all of Europe, "the temperate zones of the Americas," Australia, and the southern tip of Africa. This region is to be purged of all nonwhites. The group also calls for the creation of a "strong, centralized government" that is "wholly committed to the service of [the white] race and subject to no non-Aryan influence." These ideals of authoritarianism and *lebensraum* reflect the degree to which the NA's ideology has incorporated Nazi ideology. So does the group's adherence to biological determinism, hierarchical organization, rhetoric that emphasizes will and sacrifice, and support for "a long-term eugenics program involving at least the entire populations of Europe and America."

NA propaganda dehumanizes both Blacks and Jews, depicting them as threats to "Aryan culture" and "racial purity." However, Jews are considered a far greater and more urgent menace to white survival. In his essay "Who Rules America?" Pierce wrote, "The Jewish control of the American mass media is the single most important fact of life, not just in America, but in the world today. There is nothing—plague, famine, economic collapse, even nuclear war—more dangerous to the future of our people."

Pierce's authorship of *The Turner Diaries* brought him renown in far-right circles throughout the world. As we've seen, the *Diaries* are thought to be the inspiration behind the Oklahoma City bombing and the crime spree in the early 1980s perpetrated by the Order. More recently, members of a white supremacist gang calling itself the Aryan Republican Army committed twenty-two bank robberies and bombings across the Midwest using tactics reminiscent of the Order. *The Turner Diaries* was required reading for the group.

In March 1998 federal authorities arrested three members of a group of white supremacists in East St. Louis, Illinois, who called themselves the New Order. They allegedly planned to bomb the offices of the Anti-Defamation League in New York, the Southern Poverty Law Center in Montgomery, Alabama, and the Simon Wiesenthal Center in Los Angeles. It has also been reported that *The Turner Diaries* was a great influence on David Copeland, a British neo-Nazi who set off bombs in ethnic neighborhoods and a gay bar in London, killing three people in April 2000.

A number of NA members have been implicated in violent crimes. In June 2001, for example, NA member Eric Hanson was killed in a shoot-out after resisting the Illinois state police's attempt to arrest him on weapons charges. Hanson, who in 1999 had been convicted for physically threatening an interracial couple and for possessing illegal weapons in two separate cases, seriously wounded one of the officers who tried to arrest him.

The NA has recruited members among United States Army personnel at Fort Bragg in Fayetteville, North Carolina—with deadly results. A member of the elite 82nd Airborne Division, Robert Hunt, reportedly worked as an NA recruiter while stationed at Fort Bragg. In April 1995, according to the group, Hunt rented a billboard outside the base and used it to post an NA advertisement and local phone number. Several months later, in

December 1995, a black couple was gunned down nearby in what prosecutors called a racially motivated killing. James Burmeister and Malcolm Wright, members of the 82nd Airborne, were convicted of the murders and sentenced to life in prison. Burmeister and Wright were active neo-Nazi skinheads and reportedly read National Alliance propaganda.

Another racial killing linked to National Alliance propaganda occurred in April 1996, in Jackson, Mississippi. Police say Larry Wayne Shoemake piled a small arsenal of weapons into an abandoned restaurant in a predominantly Black neighborhood and, from this hideout, began shooting wildly onto the street, killing one man and injuring seven others. Shoemake ultimately took his own life as well.

In a police search of Shoemake's home, authorities found Nazi material as well as literature from the National Alliance. According to his former wife, Shoemake first encountered NA propaganda in the mid-1980s, when he borrowed *The Turner Diaries* from a friend, and he was not the same after reading Pierce's novel. "It was like an eye-opener to him," she said. Shoemake had also begun subscribing to Pierce's monthly publications.

Pierce's follow-up to the *Diaries,* called *Hunter,* is also popular among white supremacists. It tells the story of a racist serial killer who tries to cleanse America of its "sickness" by murdering interracial couples, eventually "working his way up" to assassinating Jews.

Pierce was deeply interested in building ties to white nationalists abroad. Over the years he built close ties between the National Alliance and the British National Party, a racist, anti-minority, neofascist party in Great Britain, and with the National-demokratische Partei Deutschlands (NPD, or the German National Democratic Party), an ultra-right-wing nationalist party in Germany. (After a series of violently racist and anti-Semitic incidents in Germany in 2000, authorities there have proposed banning the NPD, which attracts a large neo-Nazi following.) Pierce made a number of trips to Germany to attend NPD events and also invited NPD members to his headquarters in West Virginia.

The NA's interest in building ties with the far right abroad goes beyond visits and exchanges of information: the group has attracted members from Great Britain, Holland, France, Germany, Scandinavia, South Africa, Russia, the Czech Republic, and Canada. The group also works with other extremist organizations in the United States. In 2000 NA members attended and participated in events with members of David Duke's racist outfit NOFEAR (National Organization for European American Rights), recently renamed EURO, the European-American Unity and Rights Organization; the Council of Conservative Citizens; and the neo-Nazi American Friends of the British National Party. Recently, *Resistance* magazine featured an interview with Matt Hale, leader of the World Church of the Creator, as well as an article written by Hale. (In January 2003 Hale was charged with obstruction of justice and solicitation to murder a federal judge, leaving his "church" seemingly rudderless and moribund, at least for the time being.)

In addition, National Alliance units have sponsored rallies featuring former Klan leader and NOFEAR founder David Duke, and speeches by David Irving, the British Holocaust denier.

In 1999 Pierce and the NA developed a new means of communicating with possible young recruits. In April of that year he purchased Resistance Records, a music company that features white power bands playing a mix of folk, rock, Oi, and heavy metal music, with fierce lyrics directed against Jews and other minorities. Resistance was originally founded in 1993 by neo-Nazi skinheads from Canada who operated the company out of Detroit in order to avoid Canada's strict antihate propaganda laws. In addition to selling CDs, the company published *Resistance* magazine, which featured articles on the white power music scene.

American and Canadian authorities investigating tax evasion charges and violations of Canada's antihate propaganda laws raided Resistance in 1997; the company was temporarily put out of business until it was purchased in 1998 by Pierce's old nemesis, Willis Carto, and Carto's business partner, Todd Blodgett, a former low-level staffer in the Reagan administration. Pierce and Blodgett later worked out the deal that led to Pierce's taking control of Resistance.

In the fall of 1999 Pierce also purchased Nordland Records, a Swedish white power music company, and folded it into Resistance Records, in effect doubling the company's inventory. In addition to selling CDs, Pierce relaunched *Resistance* magazine, which had ceased production since the raid on the company.

Pierce appointed Erich Gliebe, who headed the NA unit in Cleveland, Ohio, to manage and promote Resistance Records. In the winter 2000 issue of *Resistance* Gliebe discussed the effectiveness of white power music in "awakening and mobilizing the White Youth of today into a revolutionary force to destroy the system." In addition to Gliebe, other NA members have taken an active role in producing the magazine.

For Gliebe, hate rock—or, as he calls it, "resistance music"—is not fun and entertainment but instead a way to further the white race. "The direction that must be taken today is of a much more serious nature, with less emphasis on drinking, fashion, and 'movement' hobbyism, and with more weight given to education, serious political activism, and total dedication to the Cause," he asserts. According to Gliebe, "complete and total White victory at any cost, then, is the ultimate goal, and resistance music is one of the weapons we are using to secure that goal."

Particularly troubling is Gliebe's dedication to recruiting teenagers, whom he recognizes as "rebellious by nature" and seeking to increase their self-esteem. "Much of our energy must . . . be directed at White youth who are not yet old enough to vote, or perhaps even drive," he admits, adding, "It is important that we reach these kids before they go off to college and are really hit hard with Jewish, multiracial propaganda; before they're old enough to go to the bars and night clubs and fall in line with the lemmings; and, of course, before they start race-mixing or decide to 'experiment' with homosexuality."

To reach teens, Gliebe says he attends "as many mainstream Rock concerts as possible," distributing literature and making contacts. He claims to have also distributed NA literature at high school sporting events, which he describes as "fertile recruiting grounds."

What kinds of message is Gliebe peddling to youth? It's a toxic blend of homophobia, racism, anti-Semitism, and xenophobia, all set to a hard rock beat. Here are a few samples drawn from the lyrics of popular white power groups:

> Stop the threat of AIDS today
> Cripple, maim or kill a gay
> We've got to take a stand today
> We've got to wage a war on gays.
> —MIDTOWN BOOTBOYS

> You say you've seen the Holocaust,
> You ain't seen nothing yet.
> Six million lies will not compare
> To what you're gonna get.
> We'll destroy the grip your money holds
> On leaders of this nation.
> Only your extinction guarantees
> The White salvation.
> —MUDOVEN

Die Jew. I hate you
You are nothing but a fool
Line you up, cut you down
Where you belong is in the ground
Choke you hard, slit your throat
Kick you in the teeth
Break your back, hear that crack
I think I'll break your feet
—INTIMIDATION ONE

Rock and roll has always been a symbol of youth culture and experimentation. But the white power music movement subverts that spirit in the service of violence and racism, subtly sending teenagers a message that bigotry is just another form of antiestablishment rebellion.

The National Alliance's forays into the "hatecore" music business are part of a well-considered attempt to advance the NA's agenda in the United States and in Europe—and they are also potentially lucrative. The National Alliance stands to make millions of dollars as the white power music industry grows in Europe and the United States.

Following Pierce's death in July 2002, Gliebe succeeded him as chairman. Although Gliebe's stewardship of Resistance Records suggests a strong entrepreneurial bent, he is every bit as radical in his views as Pierce.

For example, Gliebe has praised the Islamic extremists who attacked America on September 11, 2001, saying they "were serious, patient, and organized, and they had the discipline to keep their mouths shut so as not to leak any information about what they were planning."

Under Gliebe's continuing leadership, Resistance Records is exploring new ways of spreading hate. On Martin Luther King Day in 2002 Resistance began to advertise *Ethnic Cleansing*, a

CD-ROM-based computer game whose object is to kill "subhumans"—that is, Blacks and Latinos—and their "masters," the Jews, who are portrayed as the personification of evil. The ads for the game said, "Celebrate Martin Luther King Day with a virtual Race War!"

Patterned after popular mainstream video games such as *Quake* and *Doom, Ethnic Cleansing* has as its premise that a city—clearly New York—has been destroyed by gangs of "subhumans" controlled by Jews who are led by the "end boss" lurking in the subterranean "Lair of the Beast." Plans for world domination are seen in the subway, along with a map of "problem" areas in the United States and a sign reading "Diversity, It's Good for Jews."

The player (who can choose to dress in KKK robes or as a skinhead) roams the streets and subways murdering "predatory subhumans" and their Jewish "masters," thereby "saving" the white world. During the game, monkey and ape sounds are heard when Blacks are killed, poncho-wearing Latinos say, "I'll take a siesta now!" or "Ay carumba!" while "Oy vey!" rings out when Jewish characters are killed. The game has a high level of background detail, and various National Alliance signs and posters appear throughout while racist rock blares on the sound track.

At the end the player confronts the "end boss," a rocket-launcher-wielding Ariel Sharon, who hurls insults such as "Oy vey! Can you shoot no better than that?"; "We have destroyed your culture!"; and "We silenced Henry Ford." When Sharon dies, he coughs out, "Filthy white dog, you have destroyed thousands of years of planning."

The National Alliance is advertising *Ethnic Cleansing* as the first in a series of games to be produced by Resistance Records. The next release will be *Turner Diaries: The Game,* based on *The Turner Diaries.* While *Ethnic Cleansing* is the most sophisticated racist game available online, it's not unique. Gary (Gerhard) Lauck of Lincoln,

Nebraska (also known as the Farmbelt Fuehrer), has several anti-Semitic "entertainment" video games on his Web site under the heading "Nazi Computer-Spiele" or "Nazi Computer Games."

Most of these games are much simpler than *Ethnic Cleansing*, but they serve a similar purpose in that they allow players to interact in a racist environment in which they can indulge their fantasies. In addition, the "comedy" section of the Web site of the racist, anti-Semitic World Church of the Creator includes download-able racist games such as *Aryan 3, Shoot the Blacks, NSDoom* (NS is short for National Socialist), and *WPDoom* (WP stands for White Power).

The National Alliance is growing today at a time when other neo-Nazi organizations are becoming weaker and more frag-mented. Moreover, the NA does not appear to be siphoning members from these declining groups but actually recruiting a fresh cast of educated, middle-class bigots and young, alienated racists. These new followers appear to be attracted to the National Alliance's dedicated membership, its commanding pres-ence on the Internet, its powerful leadership, and its reputation and history, all of which distinguish the organization from other hate groups.

BIGOTRY AS POLITICS:
THE COUNCIL OF CONSERVATIVE CITIZENS

By comparison to the National Alliance and other traditional ex-tremist groups, the Council of Conservative Citizens (CCC) may appear to be a more mainstream, even benign organization. But the polished surface of the CCC hides a dangerous underbelly. The organization specializes in inflaming fears and resentments, particu-larly among Southern whites, with regard to Black-on-white crime, nonwhite immigration, attacks on the Confederate flag, and other issues related to "traditional" Southern culture. Although CCC's leadership claims that the group is not racist, its publications, Web sites, and actions all promote the purportedly innate superiority of white people and bias against nonwhites, propounding bigotry in the guise of hot-button conservative advocacy on such contentious issues as immigration, gun control, and affirmative action.

What's new and alarming about the CCC is the ease with which it has won support from many mainstream figures, lending a veneer of legitimacy to its bigoted viewpoint. The CCC drew national attention in 1998 when it was revealed that then Senate majority leader Trent Lott was a frequent speaker at its events. Subsequent news accounts reported that many other prominent elected officials had appeared at the group's gatherings.

The roots of the CCC rest in white opposition to integration during the civil rights movement of the 1950s and 1960s. The group is a successor to the Citizens' Councils of America (originally configured as the White Citizens' Councils), an overtly racist organization formed in the 1950s in reaction to the Supreme Court's *Brown v. Board of Education* decision outlawing school segregation. Trumpeting the "Southern way of life," the CCA used a traditionalist rhetoric that appealed to better-mannered, more discreet racists; while the Klan burned crosses, the CCA relied on political and economic pressure.

The organization grew quickly, attracting members from across the South and beyond; by August 1955 Patterson's membership list exceeded sixty thousand people and included 253 councils. However, as African Americans began to win greater civil rights during the 1960s and 1970s and become more politically active and influential, Southern states and their elected officials gradually liberalized. Seemingly defeated, the CCA movement sharply declined, becoming moribund by the late 1970s. But many members of the CCA retained their racist views after the organization's decline, a circumstance that would eventually allow for its rebirth.

In 1985 thirty men met in Atlanta, Georgia, among them Robert Patterson; Saint Louis attorney and former CCA Midwest field organizer Gordon Lee Baum; and William Lord, another former CCA organizer. Brought together by their frustration with government "giveaway programs, special preferences and quotas, crack-related crime and single mothers and third-generation welfare mothers dependent on government checks and food stamps," they saw the opportunity to renovate the Citizens' Council movement.

Using old Citizens' Council mailing lists, they established a new organization, the Council of Conservative Citizens, and named Baum as chief executive. The group rapidly gained adherents—including many former CCA members—and by 1999, according to the Southern Poverty Law Center, numbered 15,000 members in more than twenty states. CCC has been particularly active in Mississippi, Alabama, and Georgia.

The beliefs of the CCC fall within the racially charged tradition of its predecessor but reflect the contemporary fears of its constituency. Instead of segregation, CCC members focus on issues like interracial marriage, which the group calls "mongrelization of the races"; Black-on-white violence; and the demise of white Southern pride and culture, best exemplified in the debate about the Confederate flag. Additionally, in its heightened rhetoric about the expropriation of states' rights by the federal government and by an impending "new world order," the CCC shares some of the conspiratorial fears of modern militia groups and other right-wing conspiracy theorists.

Both on its national and chapter Web sites and in its primary publication, the *Citizens Informer,* CCC's belief in white superiority and its derision of nonwhites, particularly African Americans, are delineated without apology. The Web site of the Arkansas chapter, for instance, elaborates on the toll that interracial marriage ostensibly takes on "European American culture": it is totally unacceptable, the site states, "to think that we should voluntarily commit cultural and racial abdication."

On the group's California site, contributor Peter Anthony states, "Just as breeds of dogs are different, races of people are different as well. And just as no two cultures created by different races are even remotely alike, no two races have the same destiny in the eyes of God." Anthony also waxes nostalgic for the bygone era "when the Klan could 'march on Washington' to the cheers of an adoring public, when race-mixing and homosexuality were taboo, when racial separation was the norm."

The Confederate flag's deeply rooted connection to white Southern pride and identity—shared throughout the white South—has made it a powerful rallying point for the CCC since the early 1990s, and the controversy over its display has been a useful recruiting tool, attracting both conservatives and extremists to the organization. The main CCC Web site frames the issue as one of protecting and defending "Southern Heritage" and includes announcements for all upcoming protests as well as links to articles addressing the various debates across the South about the flag's display. A November 1999 article in the *Augusta Chronicle* (Georgia) noted that the CCC had protested at the state capital fifteen times, and while unsuccessful there, its aims were recently realized in Mississippi, where more than 60 percent of voters chose to retain the Confederate symbol in their state flag.

When the NAACP urged African Americans in particular not to vacation in South Carolina until the Confederate flag was removed from the statehouse, the CCC encouraged its constituents to take advantage of the absence of African American tourists; it posted a flier pronouncing, "Now that the African

Americans are boycotting South Carolina over the Confederate Flag, Whites can enjoy a civil liberty that has been denied to them for many years at hotels, restaurants, and beaches: the freedom to associate with just one's own people."

The views of some CCC members go even further. In an article posted on the Arkansas Web site, for instance, Dr. James Owens, a former dean of the American University School of Business, hypothesizes that a second civil war is imminent and suggests that Southern states secede from the Union in hope of creating segregated living spaces for the country's different races. In his scenario the "silent, white majority" will become shocked into taking action by the catastrophic genetic effects of interracial marriage and by the inevitable rise of an accompanying police state; a white rights movement will be forged across the political continuum, "ranging from moderate political activists . . . to the overt hostility of white supremacists and militias." Individual differences with regard to tactics will be suppressed, and the "white preservation party" will succeed in elevating and arousing "white consciousness to action" and restoring the country to "its original Euro-white dominance."

There are other suggestions that many CCC members may be more radical than the organization's public face indicates. Openly white-supremacist organizations advertise in the *Citizens Informer,* including the T. C. Allen Company, which sells pamphlets arguing that integration leads to genocide and that the biblical Adam was father of only the white race; the Ohio-based Heritage Lost Ministries, a racist and Third Position organization known also to distribute National Socialist Movement literature; and the *Resister,* the racist and anti-Semitic "political warfare journal of the Special Forces Underground."

A similar racist ideology is echoed in the editorial columns of the *Citizens Informer,* although in a somewhat muted form. Many articles consist of either tributes to the superiority of the white race or diatribes about Black violence or Hispanic immigration. As Robert Patterson, the publication's past editor, has written in a column, "Any effort to destroy the race by a mixture of black blood is an effort to destroy Western civilization itself." Columnist H. Millard offered a similar observation and a more visceral anxiety about intermarriage when he argued that minorities are turning the United States population into a "slimy brown mass of glop." Other essays in the publication lament the victimization of whites at the hands of minorities and the liberal "elite."

Other contributing writers to the *Citizens Informer* have included Jared Taylor, publisher of *American Renaissance,* which argues that African Americans are genetically inferior; Indianapolis Baptist Temple pastor Greg Dixon, who believes that churches are not bound by human laws or regulations; and psychology professor Glayde Whitney, who wrote the preface to David Duke's racist and anti-Semitic "autobiographical thesis" *My Awakening* ("Completely separately from David Duke," Whitney wrote, "my inquiries led to essentially the same places and some of the same conclusions that he spells out in this book.")

Not surprisingly, anti-Semitism is part of the CCC's cocktail of doctrines. Speakers at CCC conferences have included such figures as Edward Fields, publisher of the *Truth at Last,* a newsletter featuring articles about Jewish control of the media and government; Edward Butler, whose newsletter, the *New World Today,* describes the United States as "the Zionist slave state"; and Michael Collins Piper, a correspondent for the *Spotlight,* a newspaper published by the virulently anti-Semitic Liberty Lobby.

Despite its racist ties, the CCC has attracted the attention and support of seemingly mainstream conservative leaders. The group was catapulted to national prominence in December 1998 when a *Washington Post* reporter revealed that earlier in the year then Georgia Congressman Bob Barr had spoken before its national board in Charleston, South Carolina. Barr denied being aware of the CCC's racist views, claiming that the information packet he had received from the organization had not revealed these positions, which he considered abhorrent.

However, the ante was upped a few days later when *Post* reporter Thomas Edsall revealed that then Senate majority leader Trent Lott had appeared as the keynote speaker at a 1992 meeting of the CCC in Greenwood, Mississippi. In the article Edsall cited an issue of the *Citizens Informer* that featured a large photograph of Senator Lott at a CCC conference and quoted him as telling attendees that "we need more meetings like this." According to the *Informer,* Lott asserted, "The people in this room stand for the right principles and the right philosophy. Let's take it in the right direction, and our children will be the beneficiaries."

Lott's involvement was more complicated and arguably more deserving of condemnation than Barr's. For one thing, Lott originally denied firsthand knowledge of the CCC and later had to backtrack on this statement. For another, unlike with Barr, there was evidence of an ongoing relationship between Lott and the CCC. Leaders of the organization revealed that Lott had spoken to them on more than one occasion; that his syndicated column regularly ran in the *Citizens Informer;* and that his uncle, a member of the council's executive board, called him an "honorary member." Lott later criticized the CCC's use of his name in their publications, denied being a member, and claimed ignorance as to the racist nature of the group's rhetoric. However, Lott's claims of ignorance struck many observers as disingenuous, especially in view of his earlier public praise of the organization and its philosophy.

Edsall's reporting revealed that Lott and Barr were only the most prominent of a number of conservative politicians who had developed ties of varying intimacy with the CCC, most notably former Mississippi governor Kirk Fordice, who not only attended CCC meetings but was quick to defend the CCC in the press as well. Additionally, a Mississippi CCC leader boasted that thirty-four members of the Mississippi legislature counted themselves among the ranks of the five thousand individuals claiming membership in the state.

Other prominent mainstream political figures who have attended CCC meetings or addressed the group include former Alabama governor Guy Hunt, United States senator Jesse Helms, United States representative Mel Hancock, Alabama public service commissioner George C. Wallace Jr., Tennessee Republican national committeewoman Alice Algood, South Carolina Republican national committeeman Buddy Witherspoon, and former Arkansas supreme court justice Jim Johnson. Others who have attended CCC meetings include media figures such as editorial cartoonist Michael P. Ramirez, Accuracy in Media head Reed Irvine, and Joseph Sobran, a syndicated columnist and former senior editor for the *National Review*, whose anti-Jewish bias contributed to his firing by that magazine.

Above all, however, the appearances of Barr and especially Lott elicited widespread media coverage. In response to the revelations, Representatives Robert Wexler, a Florida Democrat, and Michael Forbes, a New York Republican, introduced to Congress in January 1999 a resolution that condemned the racism and bigotry espoused by the Council of Conservative Citizens. The resolution, modeled after a 1994 House resolution criticizing former Nation of Islam member Khalid Muhammad for racist and anti-Semitic remarks, also condemned manifestations and expressions of racial and religious intolerance wherever they occurred. But whereas the resolution against Muhammad passed through both houses of Congress in twenty days, the criticism of the CCC never even made it to the floor, largely because of the reluctance of Republicans to accept what amounted to an indirect censure of their leadership.

Instead, Representative J. C. Watts of Oklahoma, who was then the only African American Republican in the House, introduced a resolution that condemned racism in general. Watts's resolution failed. This ended the CCC episode in Congress—at least until December 2002, when Lott made remarks that seemingly defended the practice of racial segregation at a birthday party for Senator Strom Thurmond. The resulting furor led to Lott's resignation as Senate majority leader and revived publicity about Lott's ties to the CCC.

Shockingly, it remains an open question whether the United States Congress in 2003 is willing to openly condemn their colleagues who espouse racism, anti-Semitism, and bigotry.

Bringing Bigotry into the Mainstream

As the story of the CCC illustrates, there is no hard-and-fast line separating hard-core racist and anti-Semitic groups from mainstream America. Rather, there is a continuum extending from the most extreme groups, which openly espouse violence, to more "respectable" groups like the CCC, which condone bigotry in the guise of conservatism.

Thus, through channels involving right-wing circles and certain fundamentalist religious groups, bigotry and bias have increasingly infiltrated mainstream American opinion over the past two decades. It's a trend that has only been exacerbated by the heightened atmosphere of tension, fear, and intolerance that is permeating our society since September 11.

In the next chapter we'll examine how some of America's most popular and influential religious leaders have flirted with anti-Semitism and in some cases clearly crossed the line into out-and-out advocacy of bigotry and hatred.

CHAPTER 5

Jewish Calves and Christian Lions:
Dissecting the Politics of the Religious Right

AS WE'VE SEEN, the extreme right in America combines a variety of impulses—political, social, racial—in a frightening and toxic mixture. One of the most disturbing elements of this recipe for hatred is its religious component. Groups and movements such as Aryan Nations, Christian Identity, and the Phineas Priesthood brazenly commandeer religious symbols, texts, and arguments in support of their bigoted and hateful philosophies. In the process, they have done much to besmirch religion itself in the eyes of many, who wonder how the Christian God of compassion and reconciliation has been transformed into a bloodthirsty deity calling for the destruction of entire peoples.

All Christians of goodwill understand that the true message of the Gospels is one of love, not hate. Sadly, that message is now too often obscured—and not only by the hatemongers of the lunatic fringe. To a significant degree, the very word *Christian* has been appropriated by one element in the faith and yoked to a particular political agenda that bears, at most, a highly questionable connection to the teachings of Jesus. I'm referring, of course, to the use of the word *Christian* to characterize the right-wing views of some in the conservative, fundamentalist branches of American Protestantism.

These conservative Christians, whom I'll call, for simplicity, the religious right, are not unique among religious groups in their political activism. Throughout American history religion has helped drive a variety of social movements, from abolition and temperance to civil rights and the peace movement. And in many ways this has been a positive factor in our history. Like all citizens, Christians have a right and a duty to organize, lobby, demonstrate, and educate on behalf of their views. And of course the political views of all people of faith are inevitably affected by their religious beliefs.

What's different about today's religious right is the degree to which they seek to impose their personal religious beliefs and values on all Americans, equating their own brand of fundamentalist Protestantism with all faith and even with moral decency in general. They bring to political controversies the absolute certainty of the zealot, an attitude that inevitably undermines the traditions of tolerance and mutual respect on which American liberty is founded.

In this chapter I'll explain why the ascendancy of the religious right worries me and so many other Americans—and why we believe that the defense of freedom and the Constitution requires us to challenge many of the positions that the religious right has taken.

WHAT IS THE RELIGIOUS RIGHT?

Let's begin with some definitions.

Protestantism in America is highly diverse. It includes the so-called mainstream denominations, such as the Episcopal, Methodist, Lutheran, and Presbyterian churches, some of which are in turn subdivided into several distinct organizations. Taken together, these mainstream churches boast over forty million members, and they probably represent what many Americans think of as traditional Christian teaching and practice.

I don't currently see any major threat of anti-Semitic activity growing out of the mainstream Protestant churches. However, not all mainstream Protestants support religious pluralism and tolerance. For example, take the Evangelical Lutheran Church in America (ELCA), with over five million baptized members the largest Lutheran body in this country. It wasn't until April 1994 that the ELCA formally repudiated the anti-Semitic writings of its historic founder, Martin Luther—and even so, the group found it necessary at the same time to publicly "forgive" the Jews for their supposed crime of murdering Jesus. Clearly this is a denomination whose leaders still have much to learn—about history, about theology, and about religious tolerance.

The fastest-growing branch of Protestantism includes a number of churches with a strong evangelical or fundamentalist strain. (*Evangelical* refers to their commitment to proselytizing, actively seeking converts to their own beliefs; *fundamentalist* means regarding the Christian Bible as the unerring word of God, containing all that is needed for salvation and the good life.) The Southern Baptist Convention, with over thirty million members, is the largest fundamentalist denomination in the United States. It is these evangelical and fundamentalist Christians who make up the bulk of what many call the religious right.

The relationship of the Jewish community to evangelical Protestants is rather complicated. Many Jews have some values in common with evangelical Christians. Like other Americans, many Jews share their concern over changing standards of morality as well as their desire to see religious belief treated with respect in popular culture. And many evangelical Protestants are strong supporters of the state of Israel, which as we've seen is an extremely important issue for most American Jews.

At the same time, many of the positions espoused by certain evangelical Protestants trouble us—and a few are downright scary.

Some of the worst anti-Semitic attitudes that grow out of discredited Christian theologies survive among evangelical Christians. These attitudes relate to the close link between anti-Semitism and particularist views on religion. Particularism is the tendency to believe that one's own faith is the only valid path in life. A particularist is apt to think, "I have the truth, and everybody else is wrong." It's not much of a leap from this kind of thinking to hatred of those who belong to other faiths. After all, if you think God intends to subject those others to eternal damnation, surely they must deserve it.

Among Christians, particularists are inclined to accept the old theological canards that have been used to justify anti-Semitism, for example, the belief that the Jews rejected and killed Jesus. As we've seen, the Inquisition, the European expulsions of the Jews, and the pogroms were all bolstered by these beliefs.

Particularism has a long history in America. As everyone learns in school, our country was founded largely by European refugees in search of religious freedom. But people often forget that the colonists often practiced tolerance only for particular sects, not for all. The Pilgrims, for example, settled in Massachusetts because they wanted freedom to practice their faith, which had been suppressed in their native England; but they did not give the same freedom to others. (Roger Williams, a left-wing Puritan leader, established Rhode Island specifically in order to provide a setting in which all religions could be freely practiced, which was not the case elsewhere in New England.) In fact, the history of American religion can be viewed as a long struggle, still continuing, between the forces of particularism and those of tolerance and liberty.

Particularism continues to have its power. Many were shocked in 2001 by the suspension of the Reverend David Benke, a much-admired Lutheran pastor, for the offense of praying with non-Christians at a service at Yankee Stadium in New York memorializing the September 11 attacks. Benke belongs to the Lutheran Church—Missouri Synod, which is known for its conservative views. Nonetheless, this action by the denomination seemed a surprisingly harsh response to a gesture that was intended to symbolize reconciliation and national unity in the wake of an awful act of violence.

I've been wrestling with some Christians' disdain for Judaism for a long time. Back in August 1980 the Reverend Bailey Smith, then president of the Southern Baptist Convention, told a gathering of fifteen thousand at a national affairs briefing in Dallas, Texas, "God does not hear the prayers of Jews." (It's typical of a particularist to claim that he knows the thoughts, intentions, and judgments of God.)

Naturally, Smith's remark provoked a barrage of protests. After some halfhearted, awkward attempts to quell the storm, Smith visited the ADL offices in New York City. His comments to me were quite revealing. Smith said that until this controversy he had never realized that there was anti-Semitism in the United States! And what had opened his eyes? The fact that, as soon as the story hit the papers, he began receiving mail from "Christian" supporters saying, in effect, "Thanks for telling it like it is! You really gave it to those Jews."

Smith professed to be appalled by these hateful sentiments. "I need your help," he told us. "I want to stand with you, against bigotry. Can you help me figure out how to respond, and undo the damage I've done?"

We talked for a couple of hours, finally joining hands in silent prayer. It seemed like a real breakthrough moment. But a couple of years later Smith backtracked. In a public statement he said he'd been "forced" to back away from his original statement by pressure from the Jews.

An ongoing theme of particularist theology is the notion that Christians ought to devote their energies to "saving" the Jews—that is, to converting them into Christians. Although it is supposedly motivated by love for the Jews, this idea is inherently anti-Semitic in that it implicitly denigrates the value of Jewish belief. In recent years the Catholic Church has formally renounced efforts at converting the Jews. But America's single largest Protestant body, the Southern Baptist Convention, has moved in the opposite direction. In a resolution passed by their national convention on June 14, 1996, the Southern Baptists called for continuing attempts to convert the Jews; then, in September 1999, their convention specifically called for prayers that Jews would convert during the High Holy Days that fall.

In a bizarre twist, the Southern Baptists have even attacked the Catholic position as anti-Semitic. According to their peculiar worldview, since only a person who has accepted Jesus as his or her savior can escape eternal damnation, to cease attempts to convert the Jews is tantamount to condemning them to hellfire.

A CHRISTIAN NATION?

The particularist tendencies of the evangelical Protestants are especially worrisome because some of them are all too willing to carry their triumphalist attitudes into the public arena. Many on the religious right have no compunctions about proclaiming the United States a Christian nation, with the clear implication that all non-Christians—not only Jews but Muslims, Hindus, atheists, agnostics, and millions of others—are somehow not real Americans. And those in the evangelical leadership who aspire to political influence would apparently like to see the power of the state used to turn this exclusionary vision of America into a practical reality.

When it comes to this issue, today's conservative movement is actually the opposite of conservative. Rather than maintaining and protecting the tradition of separation of church and state, which has served our nation well since the days of our founders, many of them seem ready to destroy that tradition in the interests of a sectarian agenda.

It's important to make my message here crystal clear. I am *not* saying that conservative Christian shouldn't exercise their right to speak out, vote, and organize on behalf of their political and social beliefs. Nor am I saying that the values inherent in religious faith shouldn't influence our politics. The danger lies not in the political activism of religious conservatives but rather in their readiness to disparage the vital concept of church-state separation and to state or imply that Christian values—as defined by them—ought to dictate government policy.

Leaders of the Christian right—most notably the prominent and influential Pat Robertson, founder of the Christian Coalition and the Christian Broadcasting Network (CBN)—have repeatedly attacked the doctrine of separation of church and state as a falsehood promulgated by "the left" and even as a "Communist" notion. (As if to cover his tracks, Robertson also, on occasion, has made passing respectful references in television interviews to church-state separation, only to reverse course when speaking before a sympathetic audience.)

The implication: if some leaders of the religious right had their way, the United States government would become officially Christian, relegating non-Christians to second-class status, with limited rights and privileges under law. This is the agenda that the religious right has pursued through a variety of means for about the past fifteen years. (The formal start of the movement might be dated to 1987, when Pat Robertson launched his first campaign for the presidency, or to 1989, when he founded the Christian Coalition.)

During this period the religious right has operated on many fronts. Evangelical Christians have supported efforts to overturn court rulings and rewrite laws to introduce official, sectarian prayer into public schools. They've promoted candidates for local school boards, sometimes using stealth tactics to disguise their intentions, with the goal of eliminating evolution, sex education, and other "anti-Christian" subjects from public high schools. They've looked for opportunities to restrict abortion rights and to make it more difficult for women to exercise those rights. They've sought tax dollars to support religious schools and sectarian social service agencies. They've even worked to alter the federal tax code to reward stay-at-home moms and penalize working women.

You may agree or disagree with some of these policy positions. That's not the point. The point is that many leaders of the religious right have adopted these positions as part of a broader strategy to transform American government into a wholly owned subsidiary of the evangelical movement—a concept that is utterly alien to the constitutional vision of the founders as well as opposed to the values of the vast majority of Americans.

The leaders of the religious right view the situation differently, of course. In fact, they have frequently written and spoken as if conservative Christians are the targets of bigotry and discrimination in the United States. Thus the aggressive, and sometimes deceptive, political tactics of the religious right are justified by the fact that they are in a battle for cultural survival. For example, in an October 1992 column, Pat Robertson warned his followers to "expect confrontations that will not only be unpleasant but at times physically bloody"; he asserted, "Just like Nazi Germany did to the Jews, so liberal America is now doing to the evangelical Christians. It's the same thing."

In the same vein Jerry Falwell, founder of the Moral Majority, stated in a March 1993 sermon, "Modern U.S. Supreme Courts have raped the Constitution and raped the Christian faith and raped the churches by misinterpreting what the founders had in mind in the First Amendment of the Constitution. . . . We must never allow our children to forget that this is a Christian nation. We must take back what is rightfully ours."

This is a movement that feels aggrieved because it has been forced to recognize the legitimacy of other points of view, victimized by having to share political power with people who don't accept its definitions of good and evil. But as an objective description of reality, this is absurd. It makes no sense to speak as if Christians in America aren't free to practice their faith—as if churches are being shuttered and ministers imprisoned by government fiat. Has any American politician ever experienced discrimination or been attacked for declaring his belief in Christianity? Of course not.

The reality, of course, is that those on the religious right aren't truly interested in protecting their freedom to practice and preach their own faith. That protection already exists. Instead, they seek the power to impose that faith on everyone in America, replacing pluralism and tolerance with theocracy.

CHALLENGING THE RIGHT

After observing the growing influence of the Christian right for several years, ADL became concerned enough about this phenomenon to issue a public alarm. It took the form of a book titled *The Religious Right: The Assault on Tolerance and Pluralism in America,* which we published in 1994. We expected the book to be controversial, but we didn't anticipate the furor it caused.

We tried hard to make *The Religious Right* a balanced, reasonable, fact-based study of the evangelical movement in politics. We filled the book with quotations, many lengthy, from the writings and speeches of the movement's leaders. Among the groups and people dissected in the book were the Christian Coalition and its founder, Pat Robertson; the Free Congress Foundation and its leader, Paul Weyrich; Citizens for Excellence in Education and its chief spokesman, Robert L. Simonds; the American Family Association and its leader, Donald Wildmon; and Focus on the Family, headed by James Dobson.

We did *not* accuse the leading individuals or groups in the religious right of being anti-Semitic. (We are well aware of the power of that epithet, and we're very careful about how and when we use it.) We did criticize their attempts to break down the wall of separation between church and state and their readiness to use the rhetoric of intolerance to foster their political ends.

We also showed, through their own words, how some of the most prominent leaders of the Christian right had flirted with anti-Semitism. For example, take the case of Pat Robertson. In his 1991 book *The New World Order,* Robertson traffics in such traditional anti-Semitic fare as warnings about a conspiracy of "European bankers" and the sinister dealings of various occult organizations supposedly financed by the riches of the Rothschilds:

> That same year, 1792, the headquarters of Illuminated Freemasonry moved to Frankfurt, a center controlled by the Rothschild family. It is reported that in Frankfurt, Jews for the first time were admitted to the order of Freemasons. If indeed members of the Rothschild family or their close associates were polluted by the occultism of Weishaupt's Illuminated Freemasonry, we may have discovered the link between the occult and the world of high finance. . . . New money suddenly poured into the Frankfurt lodge, and from there a well-funded plan for world revolution was carried forth.

Note, in passing, the typically waffling language of the conspiracy theorist who lacks factual evidence for the views he so desperately wants to purvey: "It is reported . . . If indeed . . . we may have discovered the link. . . ." Bringing the story up to modern times, Robertson blames this same centuries-old conspiracy for the supposed New World Order that increasingly dominates the globe: "Indeed, it may well be that men of goodwill like Woodrow Wilson, Jimmy Carter, and George Bush, who sincerely want a larger community of nations living at peace in our world, are in reality unknowingly and unwittingly carrying out the mission and mouthing the phrases of a tightly knit cabal whose goal is nothing less than a new order for the human race under the domination of Lucifer and his followers."

This is language that paranoid conspiracy theorists like Aryan Nations' William Pierce would have little trouble endorsing.

If Robertson were a fringe figure with a handful of followers, we could ignore him. That's far from the case. *The New World Order* appeared on the national best-seller list of the *New York Times;* its author founded two major cable networks, appears nightly before a television audience numbered in the millions, heads his own university, and garnered over 1.9 million votes in primary elections as a candidate for the Republican presidential nomination in 1988. That's why we felt compelled to challenge his strange vision of history and publicize its anti-Semitic roots in our 1994 book.

Before the publication of *The Religious Right*, ADL actually had a reasonably close working relationship with some right-wing evangelicals and fundamentalists. My predecessor at ADL, Nathan Perlmutter, was viewed by those on the religious right as someone who understood their viewpoint (without necessarily sharing it). So the publication of our critical study came as a shock to some of those it discussed.

The most vociferous protests came from the most prominent and powerful of the religious right leaders, Pat Robertson himself. He was on the phone to me even before the book was available, yelling, "How can you attack us like this?"

"Pat," I answered, keeping my cool, "have you read the book?" He admitted he hadn't.

"Well, why not read the book before you condemn it? I'll send you five copies by overnight mail. If you find any factual errors in it, we'll correct them publicly."

Within a couple of weeks the Christian Coalition created a forty-page analysis of our book, which they sent to us and to the press. It consisted partly of denials that they were anti-Semitic (a charge, I repeat, that we never leveled) and partly of recitations of the acts of financial and moral support that the coalition had given to the state of Israel (to which we'd given full justice in the book).

They also recruited as many Jews as they could to criticize the book, drawing on the ranks of neoconservative intellectuals. In the name of this group, they ran a half-page ad in the *New York Times* denouncing our book.

We studied their analysis and wrote our own twenty-page response, rebutting the coalition's criticisms one by one. In the end we found two (insignificant) errors in the book, which we publicly corrected and apologized for. (Being willing to admit when you're wrong gives you the freedom to speak out without fear of making a mistake. People know you're human, anyway; if you admit it honestly, they respect you all the more.)

Still, the firestorm continued. We received many letters from the supporters of the Christian right saying, in effect, "We will destroy you," and denouncing me personally as "the worst enemy of the Christian faith." It got very ugly.

I regretted these hostile reactions, but not because my feelings were hurt—I have thick skin—or because I feared for the future of ADL. What troubled me was the damage being done to Jewish-Christian relations. When tempers flare, logic and facts tend to fly out the window; needless misunderstanding and resentments are created, which may last for years, and the voices of reasonable people of goodwill are drowned out.

THE "JUDEO-CHRISTIAN" GAMBIT

A move to calm the storm was finally initiated by Rabbi Yechiel Echstein, founder and president of the International Fellowship of Christians and Jews. Rabbi Echstein offered to bring the warring parties together, and on November 29, 1994, he held a day-long conference in Washington, D.C. I attended the meeting, as did evangelical leader Jerry Falwell, Ralph Reed (representing the Christian Coalition), and a number of other Jewish and Christian representatives.

From the start the tone of the meeting was positive: honest, heated at times, but respectful. It was clear that the attendees wanted to lower the level of rhetoric and start a genuine dialogue.

One of the topics we debated was the oft-repeated evangelical claim that the United States is and has always been "a Christian nation." Someone wondered why this statement should bother me; after all, isn't it a matter of factual accuracy that the majority of Americans are in fact professing Christians?

I tried to help the group see the phrase through the eyes of a Jewish American. "Every time you say this is a Christian nation," I explained, "it raises our antennas because of the history of what Christian nations have done to Jews."

At first Jerry Falwell would have none of it. "Come on, Abe," he said, "get over it. That's not what I mean." He added, "I'm a born-again Christian. That means God has freed me from sin. So I did not inherit the past history of anti-Semitism, and I am not responsible for it."

"That may be," I replied, "but every time you use the language of exclusion, you remind me of that history—and I shudder."

Does dialogue make a difference? It can. At the end of the day Falwell was speaking in a different tone. "You know," he said, "I think I understand your feelings a little better now. It's true that I'm not responsible for past anti-Semitism. But I am responsible if the things I say stir up hurts or hatred today. From now on I intend to refer to America as a *Judeo-Christian* nation, which describes our heritage more accurately."

I'd like to report that Falwell has consistently lived up to that promise. He often does use the term *Judeo-Christian,* which is at least a little more inclusive than plain *Christian.* But he periodically forgets his promise to be more sensitive and lapses into outright bigotry, sometimes bordering on anti-Semitism. For example, in a sermon he delivered in January 1999 Falwell mused about the fundamentalist concept of the Antichrist, an evil global ruler whose appearance will signal the imminent end of the world:

> Who will the Antichrist be? I don't know. Nobody else knows. Is he alive and here today? Probably. Because when he appears during the Tribulation period, he will be a full-grown counterfeit of Christ. Of course, he'll be Jewish. Of course, he'll pretend to be Christ. And if in fact the Lord is coming soon, and he'll be an adult at the presentation of himself, he must be alive somewhere today.

Dr. Bill Leonard, a Baptist who is dean of the divinity school at Wake Forest University, offered an apt reaction: "Such is the stuff of which holocausts are made. Once you start identifying a particular religious community as the source of the most evil person in the world, what in the world have you done?" Unfortunately, Falwell stood by his remarks, denying any anti-Semitic intention.

In any case, I am not really mollified by the Christian right's attempt to identify the values they advocate as "Judeo-Christian" values or even more broadly as general religious truths. Although this gambit appears superficially inclusive, it conceals a deeper intolerance. Consider, for example, these words from a prominent political leader long associated with the religious right, Attorney General John Ashcroft. Speaking in February 2002 before the annual convention of the National Religious Broad-casters association, Ashcroft stated:

> Civilized individuals, Christians, Jews, and Muslims, all under-stand that the source of freedom and human dignity is the Creator. Governments may guard freedom. Governments don't grant freedom. All people are called to the defense of the Grantor of freedom, and the framework of freedom He created.

What is wrong with this seemingly benign description of the role of faith in politics? First of all, it equates "civilized individuals" with members of the three great Abrahamic faiths—Christianity, Judaism, and Islam. Is Ashcroft really saying that anyone who is not a member of one of these religions is not civilized?

Second, it places the political concepts of freedom and human dignity within a religious context, asserting that these must be viewed as gifts from God. The implication is that those who view God differently from Ashcroft or don't believe in God at all reject the concepts of freedom and human dignity, which would put them beyond the pale of American social discourse.

This is deeply misguided. Yes, people of faith have a role to play in national affairs—but as American citizens, not as repre-sentatives of any sect or religion. And people of faiths other than Christianity, Judaism, and Islam—Hindus, Buddhists, and many others—are equally welcome in public life, as are people of no faith at all.

Thus the continual implication by leaders of the right that only a select group of religions is truly legitimate is an offensive one.

FINDING COMMON GROUND

At the 1994 conference I got to know Ralph Reed, who came across as far more moderate and tolerant than his mentor, Pat

Robertson. In time he and I developed a positive working relationship, one that has caused some consternation among ADL friends and supporters.

At my invitation, Reed has addressed ADL forums on more than one occasion, surprising audiences with his genuine respect for the Jewish people and his willingness to live with religious, political, and social differences. Reed and I certainly don't see eye to eye on every issue, but we disagree agreeably, and we always try to resolve our differences away from the media spotlight. Both of us agree that the last thing our communities need is a public battle that seems to pit Christians against Jews.

My relationship with the controversial Reed makes many of my closest supporters very nervous. The first time I invited Reed to speak at an ADL leadership conference in Washington, my wife, Golda, worried about how I would handle myself. "Abe," she said, "I know you. You're a hugger. But if I see you hugging Ralph Reed on the TV news, don't bother coming home tonight." In deference to Golda, I maintained an arms-length policy with Reed—at least, that first time.

Today Reed runs Century Strategies, a political consulting firm, and he continues to serve as a point of communication between ADL and the religious right. In April 2002, when Israel was under siege by suicide bombers and other agents of the new *intifada*, he wrote an eloquent op-ed piece for the *Los Angeles Times* titled "We People of Faith Stand Firmly with Israel." We were so impressed with Reed's message that we reprinted the article in the *Washington Post*, the *New York Times*, and the *Washington Times*, causing a minor controversy in the Jewish community. Reed's most recent address before an ADL leadership conference was met with a standing ovation.

Well-intentioned Jews (and others) sometimes ask me, "Why would you want to legitimize a person like Ralph Reed?" My answer is connected to the larger question of how Jews should relate to the religious right. There are enormous differences between us on issues ranging from prayer in schools to gay rights. I worry about the hidden agenda of church-state melding that many of their leaders pursue. And some on the right, like Pat Robertson, have at least flirted with outright bigotry.

All these considerations make it clear that Jews ought to be wary of the motivations of some on the religious right. Furthermore, we must always maintain our willingness to criticize the leaders of the religious right when that's appropriate. And ADL and I have done just that. We've spoken out forcefully—and repeatedly—when figures such as Robertson and Falwell made insensitive statements or uttered religious slurs.

Yet there are some important areas of commonality between the Jewish community and the religious right, most notably in regard to the state of Israel. Most Christian fundamentalists are faithful supporters of Israel, a position they hold for at least three distinct reasons.

Many support Israel for the same reasons that most Americans do: because it is a democracy (the only democracy in the Middle East) and one of the three nations in the world (along with the United States and Great Britain) that is most deeply committed to the war on terrorism.

Many also support Israel because of the spiritual bond all Christians have to the Holy Land. After all, Israel is where Jesus, the founder of their faith, was born, conducted his ministry, died, and, in Christian belief, was resurrected. In modern times, access to the holy sites of Christendom in Jerusalem, Bethlehem, and other locations has been most safe and free under Israeli control (while Muslim believers continue to enjoy access to their holy sites as well). Thus a strong state of Israel represents the best hope for shared access to the Holy Land by all three of the world's great monotheistic faiths.

Many Christian fundamentalists have a third reason for supporting Israel. Based on their interpretation of specific passages from the New Testament book of Revelation, they believe that the Messiah will return and usher in his heavenly reign only after the Jews are secure in their homeland. For these believers in end-times theology, support of Israel is linked to a sense that the second coming of Christ is imminent and can even be accelerated by the actions of human beings.

Obviously, I don't share these theological presumptions. As a Jew, I am still waiting for the *first* coming of the Messiah. But I am open-minded. I figure that when the Messiah comes we can ask whether he—or she—has been here before and settle the question once and for all.

Some Jews find the fundamentalist beliefs about Israel a little weird. And as we've seen, in the hands of a Jerry Falwell these end-of-the-world doctrines can take on a distinctly anti-Semitic tinge. However, I don't consider these beliefs a reason for rejecting the support of Israel offered by the religious right. What matters to me is the depth and solidity of the support, not the reasons behind it.

I believe that for Jews to reject the political support of Israel by the religious right would make no sense, especially at a time like today when Israel is so isolated on the world scene. When you are in danger and someone offers help, you don't question the purity of their motives, you just accept the help.

My position on this matter—a controversial one in some quarters—is that, for the foreseeable future, American Jews will have to live with an influential religious right. Some of their leaders are reliable friends of Israel and the Jews, others much less so. We must manage our relationship with the Christian right with care and vigilance, criticizing the movement forthrightly when our interests and values clash and accepting its support when they

overlap. Perhaps the sage Woody Allen put it best: "The lion and the calf shall lie down together, but the calf won't get much sleep."

TODAY'S CONTINUED ASCENDANCY OF THE CHRISTIAN RIGHT

In terms of press coverage, the Christian right probably was at the apex of its fame in the mid-1990s. But its political power and social influence today may actually be greater than ever. The election of George W. Bush in 2000 brought back into Washington a Republican administration with strong ties to the religious right. Thus continued scrutiny of the most powerful forces in the movement remains vitally necessary.

The Christian Coalition remains an important player in electoral politics. The group claims nearly two million members nationwide and is affiliated with a wide array of media outlets and other institutions, including the Christian Broadcasting Network and Regent University.

The coalition publishes voter guides that are nominally nonpartisan but clearly slanted toward hard-line conservative positions. (For example, they describe candidates who support funding for the National Endowment for the Arts as favoring "tax-funded obscene art.") In fact, the partisan nature of these guides led to the group's losing its tax-exempt status in 1999. In 2000 the coalition conducted its biggest-ever election campaign, distributing seventy million voter guides through conservative churches and its own members in every state of the union.

The coalition's ties to Washington are among the strongest of any lobbying group. Its annual "Road to Victory" conference draws all of the most prominent conservative Republicans; for example, speakers in 2000 included Senate majority leader Trent Lott, House majority leader Dick Armey, speaker of the house

Dennis Hastert, majority whip Tom DeLay, Republican National Committee chairman Jim Nicholson, and presidential candidate George W. Bush (on videotape).

Pat Robertson himself continues to do mischief. For example, he has not moderated his desire to undermine the doctrine of separation of church and state in pursuit of a theocratic America. On his *700 Club* television program on October 5, 2000, Robertson remarked:

> The concept that one God, "Thou shall have no other gods before me," will somehow upset a Hindu, that's tough luck! America was founded as a Christian nation. . . . And the fact that somebody comes with what amounts to an alien religion to these shores doesn't mean that we're going to give up all of our cherished religious beliefs to accommodate a few people who happen to believe in something else.

Robertson and the institutions he heads also continue to flirt with anti-Semitic attitudes. For example, on Easter 2002 Robertson's Christian Broadcast Network (CBN) ran an animated cartoon, "The Easter Promise," showing stereotypical Jewish figures—complete with hooked noses and Yiddish accents, in first-century Palestine!—maliciously plotting the death of Jesus.

I wrote to Robertson at CBN to express my concern about this offensive cartoon. I got a very peculiar letter in reply. Robertson did not respond to my detailed complaint about the anti-Jewish imagery used in the cartoon. (Actually, he claimed never to have seen the cartoon, which one would think would preclude his defending it.) Instead, to my astonishment and dismay, he launched into a personal attack on me, beginning, "It does seem that the Democratic Party must be very concerned about the Fall elections since it sent its principal secret agent into action in

April." (The many Democrats I've publicly criticized and debated would be surprised to learn that I am a secret agent of their party.)

Robertson then went on to list some of the dire world events of the preceding year, and he concluded:

> To think that in the midst of these epic struggles, when I am doing everything in my power to defend the rights of Jews and Israel against these unprecedented attacks against them by radical Islam, the leader of the Anti-Defamation League decides he wants to pick a fight with me over a couple of cartoon characters and a few lines of fictional dialogue dealing with the ancient history of two competing groups of Israelites. I am simply appalled that you are so insular in focus.
>
> But Abe, I understand your game. It is clear that your focus is not the defense of worldwide Jewry, but the domestic political agenda of the Democratic Party of the United States and your own self promotion. In my opinion, there is no one in the entire United States who has done more to poison the relations between Jews and Christians in America than you. I marvel that the Board of the Anti-Defamation League hasn't restrained you some time ago.

It's disturbing to realize that a man who is capable of such insensitivity and defensiveness represents a large and powerful constituency of Americans.

At least Robertson is getting used to my persistence. I notice that when he makes a remark that implies stereotypical or prejudiced attitudes, he often comments, "I guess Abe Foxman will be criticizing me again." He's right—I usually do.

Another prominent force of the religious right is the Family Research Council (FRC), a think tank that lobbies for what it considers traditional family values—as defined, of course, in fundamentalist Christian terms. Founded in 1983 by the popular television minister Dr. James C. Dobson, FRC has since been headed by Gary Bauer, a former Reagan administration appointee and presidential candidate in 2000, and most recently by attorney Kenneth L. Connor, a Floridian with long experience in the antiabortion movement. The organization claims over 450,000 members, forty state affiliates, and an annual budget of over ten million dollars.

Practically all Americans believe in strong families and support government policies that encourage them. In practice, however, the FRC's efforts focus on such right-wing causes as outlawing abortion, eliminating sex education that teaches about contraception, and especially demonizing homosexuals. The FRC has even stooped to using the war on terror as an excuse for pushing this agenda:

> Do you really think that when our troops from Delta Force crawl into Osama bin Laden's cave in Afghanistan or into the face of the muzzle of a terrorist machine gun, that they are doing it so that women can kill their children, so that pornographers can peddle their smut, so that people of the same sex can marry? If those features of American life become the fixtures of American life, I fear that our nation may not long endure. ("Reflections After the Terror," October 2, 2001)

Another powerful organization linked with Dr. Dobson is Focus on the Family. Dobson is one of the most popular media voices of the religious right, with a radio and television audience that has been estimated at 200 million people per day. His political ties are strong; he held a number of advisory posts in the Reagan and first Bush administrations and was appointed to federal commissions dealing with family issues by Republican senators Robert Dole and Trent Lott.

Focus on the Family publishes ten monthly magazines with a

total of 2.3 million subscribers. Like the Family Research Council, the group pushes a narrowly defined moral agenda that emphasizes antiabortion, antigay, and anti-sex-education themes. It also works to break down the walls between religious instruction and public education, for example, by encouraging teachers to create student prayer groups.

TARGETING PUBLIC SCHOOLS

The Christian right continues to chip away at traditional notions of church and state, still aiming to make Christian faith a quasi-official national doctrine. Their efforts are especially visible in the field of education. Groups and individuals from the religious right are active in almost every state, promoting such sectarian goals as the censorship of textbooks that deviate from conservative Christian teachings; substituting the fundamentalist doctrine of creationism for the scientific theory of evolution in biology classes; and advocating organized prayer, Christian-oriented Bible studies, and displays of the Ten Commandments and other religious symbols in public schools.

They are also working heavily on behalf of school voucher programs, which they hope will channel billions of dollars of taxpayer money into private religious schools. Many groups, including ADL, oppose voucher programs in part because they drain resources from our already cash strapped public schools. This prospect doesn't trouble many on the religious right. As Jerry Falwell once remarked, "I hope I live to see the day when, as in the early days of our country, there won't be any public schools. The churches will have taken them over again and Christians will be running them." Most voucher advocates don't really want to see public education destroyed, but many of those on the religious right do—although they rarely admit it openly as Falwell did.

Other elements of the Christian right also use stealth tactics to push their educational agendas. For example, Joyce Meyer Ministries, a Missouri-based Christian evangelical church, sponsors Rage Against Destruction, a traveling musical extravaganza that puts on public school assemblies with a focus on antiviolence messages. The shows are dynamic performances featuring concert-quality, cutting-edge lighting and music and lavish giveaways to select students, including Sony PlayStations and New Balance athletic shoes. The ultimate purpose of the assemblies, however, is to invite students to an off-school event called "Firefest," an unabashed Christian evangelical festival with a high-pressure pitch aimed at vulnerable teens.

No one—not students or teachers, principals, or parents—is told about the true affiliation and purpose of the Rage Against Destruction team. Neither, apparently, does the group disclose its purpose and identity to its partners and sponsors, which, according to the Rage Web site have included New Balance, MTV, the National Education Association, and Big Brothers and Big Sisters of America. Meyer deliberately uses the public schools to promote events in which proselytizing is the main objective, typically showing no regard for the separation of church and state embodied in the Constitution. Recently, she told an audience of believers that the notion of separating church and state "is a deception from Satan."

After ADL began publicizing the real nature of Rage Against Destruction and the stealth ministry of Meyer, the organization responded by removing evidence of the evangelical program from its Web site. However, education officials are beginning to catch on. This year, several Rage assemblies in New York, New Jersey, and Missouri were canceled after state education officials warned principals about the group's tactics.

STANDING FOR INTOLERANCE

Today the national Republican Party is largely dominated by figures with strong ties to the Christian right, from Tom DeLay, Dick
Armey, Dennis Hastert, and Trent Lott to Attorney General John
Ashcroft. Votes from conservative Christians make a decisive
difference in many elections at the local, state, and national levels.
Thus President George W. Bush's readiness to cater to the religious right is politically understandable—and a source of enormous concern for me.

When Bush ran for governor of Texas against Ann Richards in
1994, an interviewer from the *Houston Post* asked both candidates
to describe their religious views. Richards declined to discuss any
beliefs that didn't affect her work as governor. But Bush elabo-
rated on his conviction that there is a heaven and a hell, adding
that "only Christians had a place in heaven."

When the inevitable furor erupted, Bush tried to defend his
statement by attributing the idea to the evangelist Billy Graham.
(Apparently he felt that the imprimatur of Graham, who was
practically the unofficial chaplain of the White House in several
presidential administrations, would win him absolution.) But
eventually he apologized for the gaffe.

Many of those in today's Republican leadership have strong ties
to the religious right. It surprised no one when House majority
whip Tom DeLay called, at an October 2002 Christian Coalition
meeting, for the election of candidates "who stand unashamedly
with Jesus Christ." The fact that President Bush is closely allied
politically with people who make such statements makes me
worry about his even-handedness and his real commitment to
church-state separation.

During the 2000 campaign I was disturbed by Bush's declaration in another interview that his favorite political philosopher
was Jesus Christ. I didn't like it any better when his opponent,

Vice President Al Gore, not to be outdone, said that he would govern by asking, "What Would Jesus Do?" I was also troubled by the overt religiosity of some of Joe Lieberman's statements as the Democratic vice presidential candidate. The injection of faith into politics always worries me, no matter who the candidate or what the faith.

The world-changing events of September 11 have not brought out the best in the religious right. Instead, they have encouraged a new wave of intolerant, divisive language in the guise of patriotism. For example, consider the infamous broadcast conversation between Jerry Falwell and Pat Robertson on the *700 Club* show of September 14, 2001:

> *Falwell:* The ACLU has got to take a lot of blame for this [the terror attacks of September 11]. And I know I'll hear from them for this, but throwing God—successfully with the help of the federal court system—throwing God out of the public square, out of the schools, the abortionists have got to bear some burden for this because God will not be mocked, and when we destroy forty million little innocent babies, we make God mad. . . . I really believe that the pagans and the abortionists and the feminists and the gays and the lesbians who are actively trying to make that an alternative lifestyle, the ACLU, People for the American Way, all of them who try to secularize America—I point the thing in their face and say you helped this happen.

> *Robertson:* I totally concur, and the problem is we've adopted that agenda at the highest levels of our government, and so we're responsible as a free society for what the top people do, and the top people, of course, is the court system.

Of course, many protested the bizarre notion that gays and feminists were somehow responsible for the terrorist attacks on

America, and Robertson soon apologized. But within a year both Robertson and Falwell were using the war on terror to justify fresh words of bigotry, this time tarring Islam as an inherently violent religion (Robertson) and calling the Prophet Muhammad "a terrorist" (Falwell).

It's terrible to see the country we love becoming the target of terrorist attacks. But if we allow the resulting feelings of vulnerability and anger to turn us into hatemongers and bigots, we will be handing the enemies of freedom an awful victory.

Unfortunately, intolerance—including flirtation with anti-Semitism—is not confined only to the right wing in America. We must battle bigotry on other fronts as well. One of the most disturbing problems we face is that of anti-Semitism in America's Black communities, an issue I'll tackle in the next chapter.

CHAPTER 6

Troubled Alliance:
The Rift Between American Blacks and Jews

THE DETERIORATION OF THE RELATIONSHIP between the African American and Jewish communities is one of the great tragedies of recent years. It's a complex problem with many causes, which can be examined in light of history and understood logically. But it's sad that these two great peoples who in the past supported one another nobly in their joint struggles against prejudice now find themselves in a relationship filled with misunderstanding, mistrust, and sometimes hostility.

Symptoms of anti-Semitism in today's African American community are numerous. They include:

· The ease with which many Black political leaders, from Jesse Jackson to Al Sharpton, have shrugged off incidents of flirtation with anti-Semitism—and the unseemly readiness of many Black citizens to ignore or excuse such incidents

· The continued prominence among Black Americans of the bigoted religious sect known as the Nation of Islam and the unmerited respect accorded its racist and anti-Semitic leader, Louis Farrakhan

· The popularity on college campuses of Black educators and organizations that espouse messages of antiwhite and anti-Semitic hatred

· The disturbingly frequent use of anti-Semitic stereotypes and epithets by popular Black entertainers from the worlds of pop music, hip-hop, and comedy

Polling data bears out my concerns. For as long as ADL has surveyed anti-Semitic attitudes, we have found that some 30 to 40 percent of Black Americans are infected with a significant degree of anti-Semitism. Forty years ago this figure was matched in the white population. That's no longer the case. Anti-Semitic beliefs and feelings have gradually declined among whites, but among Blacks they have remained high.

At a time when so many global trends are converging to isolate and threaten the security and survival of the Jews, it's disheartening to feel that the African American community may be changing from a close partner into a potential antagonist.

ALLIES AGAINST HATE

Jewish Americans have a long, proud history of partnership with Black Americans in the battle for freedom and tolerance. In fact, for much of its history, the Anti-Defamation League was almost as well known for its battles on behalf of Black Americans as for its work against anti-Semitism.

ADL helped lead the Southern movement against lynching in the 1920s and 1930s. It also helped to draft and pass legislation against wearing masks in the state of Georgia some fifty years ago. When the legislation became law in states throughout the South, it proved to be the beginning of the end for the most powerful and dangerous racist group in American history, the Ku Klux Klan. Many of the self-proclaimed heroes who advocated bigotry under the cover of a white hood suddenly disappeared when the law exposed them.

ADL also supported the civil rights movement when its greatest landmark legal case was brought: *Brown v. the Board of Education of Topeka, Kansas,* the case that established the legal principle that separate accommodations for Blacks and whites are inherently unequal and therefore legally impermissible. ADL filed an amicus curiae brief in *Brown* supporting the plaintiff.

For many years after *Brown* the battlefront in the civil rights struggle was primarily legal and legislative. Several organizations worked together to defeat Jim Crow (the infamous American version of apartheid that ruled the South and parts of the North for over two generations) and open up jobs, housing, and other public accommodations to all Americans, regardless of race. The ADL was an important leader in these battles, and more generally the Jewish community was in the forefront of the efforts to win equal rights for Black Americans.

Jews were active in support of the rights of African Americans for a variety of reasons. Altruism certainly played a role. The pursuit of justice is a powerful thread in Jewish culture, dating back to the teachings of the prophets. And the idea of a people seeking freedom from bondage is of course deeply ingrained in Jewish history and lore. It's no accident that many of the most beloved stories and gospel songs in which Black Americans have expressed their longing for liberation use imagery borrowed from the biblical story of the Jewish captivity in Egypt. American Jews saw Black Americans as brothers and sisters in search of freedom, and it was natural for them to reach out in support of Black aspirations.

There was also an element of self-interest in the Jewish support of civil rights. Our experience of anti-Semitism has made it clear that the cycle of hatred encompasses us all. The same people who hate Jews today are all too likely to hate Blacks tomorrow—and vice versa. To tolerate prejudice and hatred of any kind simply encourages attitudes and practices that someday will victimize us all. Therefore, to stand up on behalf of our Black fellow citizens was simply an extension of the principle that led us to stand up in our own defense. And our support of civil rights for African-Americans was strongly reciprocated. Black leaders spoke up powerfully against anti-Semitism and in support of the state of Israel, the liberation of Soviet Jewry, and other Jewish causes.

THE TROUBLED PARTNERSHIP

By the time I joined ADL in 1965, there was a widespread feeling that the major civil rights battles had been won. The 1964 Civil Rights Act was the cornerstone law, the great achievement that had broken down the major legal barriers preventing Blacks from claiming their rights under American law. What would follow this, we all believed, was simply a gradual process of making the promise of the Civil Rights Act into a reality through supporting legislation, legal cases to clarify the meaning and extent of the law, and educational efforts to win all Americans to willingly supporting the law and its meaning.

In retrospect, it was a simplistic, even naive, point of view. The years since 1964 have not been a time of unalloyed advancement for Black Americans. And they have certainly not been a simple extension of the good feeling between Blacks and Jews that marked the heady days of the civil rights movement. Instead, they have been a time of uncertainty, disagreement, and halting progress—two steps forward, one step back—that has led to frustration and at times despair for many African Americans and some of their white allies. It has also led to a breakdown in the formerly warm spirit of cooperation and mutual support between the Black and Jewish communities.

With the benefit of hindsight, it's easy to recognize the stages by which the Black-Jewish alliance became frayed.

First, there was the emergence of relatively radical groups alongside the older, more moderate civil rights organizations. Dissatisfied with the progress attained by organizations such as the NAACP and the Southern Christian Leadership Conference of Dr. Martin Luther King Jr., younger Blacks formed groups such as the Congress of Racial Equality (CORE), the Student Non-Violent Coordinating Committee (SNCC), and the Black Panthers. Marching under the banner of Black Power, these newly militant groups sought independence from white influence and support, taking the position that only a movement run by and for Blacks could truly achieve freedom for African Americans.

This was a positive shift, a coming of age for the African American community. But an unfortunate side effect was the creation of a breach between Black Americans and others who had been their allies. White supporters of civil rights, including Jews, began to feel unwelcome in the movement they'd help to found.

A second force that helped to create this breach was the leftward movement in the late 1960s and early 1970s of certain individuals and groups in the civil rights movement. This wasn't a shift toward the "old left" that many Jews of an older generation had supported. It was a "new left" with an aggressively anti-American slant. Many American Blacks sought a new sense of identity with Africa, their ancestral homeland, and with the so-called Third World in general. Over time these sympathies extended to many groups and nations of relatively unempowered people of color, including Arabs. And many Black Americans became interested in exploring Islam, one of the dominant religions on the African continent, as a spiritual path that might serve as an alternative to "white" Christianity.

It was a natural progression from this position to joining many Arabs in opposing the existence and security of the state of Israel. And this was a simple step for some Blacks to take, since the people of Israel were relatively wealthy (by Third World standards), largely of white, European origin, and allied with the same American establishment that was seen as supporting dictatorships in Vietnam, Latin America, and other countries around the world. Naturally, the new antipathy of many Black leaders toward Israel upset, alienated, and angered many Jews.

In addition, changes in the economic status of our two communities undoubtedly played a role in our shifting relationship. Most Americans benefited from the postwar economic boom of the 1950s through the 1970s, but Jewish Americans, as a group, did better than Black Americans. By the end of this period many Jews who had lived in city apartments in close proximity to working-class people of every ethnic background, including Blacks, had moved to homes in the suburbs.

Now many Americans Jews were physically and psychologically separated from the growing troubles of the cities: crime, drugs, poverty, and broken families. The natural closeness that had once existed between Blacks and Jews who rode the same subways, visited the same city parks and beaches, and attended the same public schools, gradually disappeared. Under the circumstances it's no wonder that the social attitudes of the two communities also began to diverge. Some Jews were influenced by their growing affluence to become more politically conservative (although statistics show that this trend is less pronounced among Jews than among other white ethnic groups in America).

These cultural and political shifts of the 1960s and 1970s encouraged the emergence of anti-Semitic feelings that had always simmered just below the surface in America's Black communities (just as they do in many white communities). The tensions that still exist between many Blacks and Jews today are a continuing outgrowth of the trends that started a generation ago.

"THE ANATOMY OF FRUSTRATION"

At its very inception, the Black-Jewish rift was analyzed by Bayard Rustin, the famed civil rights leader and director of the A. Philip Randolph Institute, in a classic address he delivered at the 55th National Commission Meeting of ADL on May 6, 1968. Titled "The Anatomy of Frustration," Rustin's speech painted a picture of life in the typical Black inner-city community that helped to explain why anti-Semitic feelings, however irrational and wrong, were springing up among Blacks.

As Rustin explained, the average poorly educated, underemployed Black American living in an inner-city community knows four kinds of white people: the police officer, the business owner, the schoolteacher, and the social worker. For various reasons, most of which are *not* the fault of the individuals in question, all four are sources of frustration, humiliation, and anger for the Black citizen. The owner of a small store in Harlem, for example, is forced to charge more for the same merchandise than a downtown department store charges—not because of racism or greed but because of the high cost of doing business in a dangerous neighborhood as well as the cost of the long-term credit that is used to finance so many purchases in the ghetto. And as Rustin pointed out, in many American cities three of those four archetypal white figures in Black communities have been predominantly Jewish (all except the police officer).

With Jews so visible in the Black community—and seemingly so entrenched in a system that oppresses Black citizens—no wonder many Blacks harbor resentments against Jews and are prone to adopt age-old anti-Semitic attitudes.

Rustin's speech was a sensitive and perceptive analysis of some of the root causes of anti-Semitism among African Americans. Much of what he had to say still applies today, although some of the specific circumstances he alluded to have changed. For example, in the late 1960s most of the schoolteachers and principals in New York City were Jewish. (My wife, Golda, was one of them. She taught school in Harlem for twenty-five years until her retirement in 2001.) Today, owing in part to improved educational opportunities for minorities and in part to the school decentralization movement of the 1970s, most educators in Black communities are Black and Hispanic.

Most of the Jewish merchants in historically Black neighborhoods are also gone. They were displaced over the past three decades, first by Blacks and other minority business owners, then by big chain stores. The coffee shop, record store, and clothing store once owned by Jewish proprietors have given way to McDonalds, Sam Goody, and Old Navy.

Despite these changes, the anti-Semitic assumptions that found fertile ground among Black Americans have lingered. And the political, economic, and social frustrations that Rustin cited as a further influence on Black attitudes continue to fester. The unhappy result has been a deep and abiding rift between two communities that had worked so closely together in the most difficult and dangerous times.

ISSUES THAT DIVIDE

The greater distance that exists between the Jewish and Black communities today as compared to the early 1960s can also be attributed, in part, to changing priorities in the Black community.

As I remarked earlier, when I joined ADL in 1965, we assumed that most of the legal and legislative battles on behalf of civil rights had been won. In fact, that was true. On paper, Black

Americans today enjoy the same legal protections as whites, and even the vestiges of Jim Crow are basically gone.

This doesn't mean that the economic and social gulf between whites and Blacks isn't still a major problem. It is. But the greatest need now is not for new civil rights legislation but for programs, policies, and social initiatives to improve the living conditions of Black Americans: to create jobs, improve education, provide health care, reduce the incidence of violence and crime, and strengthen families. Historical discrimination has played a major role in creating the ills that afflict the Black community, but simply protesting against discrimination can't fully solve them.

And when it comes to creating greater economic opportunity and a stronger social safety net, the Jewish community can't do as much to help as we did in the days when civil rights legislation and government policy were at the heart of the battle. As a practical matter, our influence in this area is relatively minor and difficult to leverage. After all, creating jobs in the Black community (for example) isn't as simple as passing a law or winning a case in court. It involves large-scale economic, social, and educational reforms that no one, white or Black, has yet figured out how to mobilize.

Some African American leaders don't agree with me on this point. In the aftermath of the terrible Crown Heights riots, I took part in many meetings with Black leaders to explore ways of healing the rift between our communities. Some challenged me to help the Black community by bringing Jewish real estate moguls to invest in Crown Heights.

As I explained, this suggestion credited me with a lot more power than I have. The businesspeople who run New York's big real estate firms—whether they are Jewish or not—are like any other businesspeople. They're looking for profitable investments. If such investments are available in Crown Heights or other Black communities, some of them will be interested. But they don't operate in lockstep, they don't pick their investments on ethnic lines—and they certainly don't look to me for business advice.

The Black community leaders who asked me to provide Jewish financial help were well-meaning. But there's a subtle element of anti-Semitism in assuming that Jews control the real estate business, or any other business, and that they act in concert to promote favored causes. I'm sure a Black leader like Al Sharpton or Congressman Charles Rangel would consider it patronizing—and even slightly racist—if I called him up and asked him to get all the Black entertainers and athletes in America to support Israel. It's very much the same thing.

Finally, another major factor in the rift between the Black and Jewish communities is the emergence, since the 1960s, of a couple of specific issues over which we have real differences of opinion. These differences can't be ignored or minimized, although they needn't be a source of hostility or hatred, either.

One of these issues is the controversial topic of affirmative action.

Matters of definition are important when discussing an issue as complex as this one. Broadly speaking, affirmative action refers to any program or policy designed to encourage and enhance the life opportunities of Blacks, Hispanics, and other members of disadvantaged groups. Affirmative action can affect hiring practices, public housing allocations, and other situations involving choice among a range of applicants, but its most important role is in college and graduate school admissions. And it is here that the major controversies have arisen.

Don't misunderstand. I believe that the impulse behind affirmative action is completely appropriate. It's obvious that minority-group members, especially Blacks, have long been denied adequate access to educational opportunities in the United States, a situation that has helped to perpetuate the economic and social disadvantages from which they suffer. It's a very defensible position to argue that America's legacy of slavery means that poor Blacks deserve—at least for a time—to be given some special opportunities to gain a foothold on the ladder of success, in partial recompense for what they suffered while helping to build our country.

It's also very defensible to contend, as many college and university administrators do, that the quality of learning improves when classrooms and campus organizations are integrated, not only racially but in terms of life experience. Students learn from one another, and they learn more when they have an opportunity to hear from people with backgrounds different from their own.

Ironically, a striking illustration of this thesis recently emerged from, of all places, today's highly conservative United States Supreme Court. In December 2002 the controversial and usually silent justice Clarence Thomas spoke up during oral arguments about the constitutionality of a state statute against cross burning. Thomas held his fellow judges riveted with an eloquent explanation of the intensely frightening, intimidating message that cross burning sends to Black Americans. By all accounts, Thomas's powerful personal perspective completely changed the tenor of the conversation, paving the way for an emphatic ruling upholding the statute. The incident underscores the value of diversity—even in an elite institution such as the Supreme Court.

So neither I nor ADL have any quarrel with the *objectives* of affirmative action. But as the saying goes, the devil is in the details. In my view, affirmative action in practice is often little more than a disguised form of quotas, a system whereby ethnic, religious, and racial groups are allocated targeted numbers of places within an organization. And I consider the use of quotas in any form to be unacceptable.

One reason for my strong dislike of quotas is that I am offended by the use of race or any similar personal characteristic as a way of categorizing people and modifying how they are treated. Over the centuries we've seen too much evil done by regimes that divided the citizenry into groups based on supposed racial characteristics, from Nazi Germany to apartheid-era South Africa. I hate to see our country drifting in that direction and away from Dr. King's vision of a color-blind America.

As a Jew, I also can't help recalling how quotas were used for decades as a way of limiting the opportunities afforded to Jews and other minority-group members in the United States as well as in some European countries. Elite colleges and universities, as well as some employers, would establish ceilings on the number of Jews they would accept. As a result, highly qualified Jews lost out on opportunities that were reserved instead for less-qualified members of other communities—in those days, mainly white Protestants. Throughout the 1940s and 1950s, ADL and other Jewish organizations fought against quotas. We are sorry to see them returning today, although the professed intentions of those instituting them are different and more defensible.

Disagreement about affirmative action became a major bone of contention between some Jews (including the leadership of ADL) and most Black leaders beginning in 1978, when the Bakke case came before the Supreme Court. Although the court ordered the University of California medical school to admit the white plaintiff in that case, Allan P. Bakke (who, as it happens, was not Jewish), it upheld the general principle of using race or ethnic origin as one factor in admission to universities or professional schools. ADL had submitted an amicus brief in support of Bakke, angering many of our traditional Black allies.

Today, after a quarter century of debate about affirmative action and a number of similar lawsuits, the future of the practice is still unclear. However, the sands of public opinion appear to be shifting, slowly but steadily, against it. Although the main established civil rights organizations, including the NAACP and the National Urban League, continue to defend affirmative action, a growing number of individual Black thinkers, including scholars and pundits such as Shelby Steele and Thomas Sowell, have come to question its efficacy.

It's possible to achieve the benefits sought by affirmative action supporters without abandoning the goal of color-blind policies. One alternative would be to award points in admissions decisions for life experiences such as poverty, disadvantage, and discrimination. This would provide a leg up to those who most need and deserve it, without differentiating among people on racial or ethnic grounds. And it would eliminate the illogical and absurd possibility of giving a special boost to (for example) the child of a Black physician from New York's Park Avenue rather than the child of a white coal miner from Tennessee.

And there are creative ways of achieving diversity on college campuses without using quotas, open or disguised. In Texas and California, admission to the prestigious state university programs is now extended to all students who graduate in the top 10 percent of their high school class. The result is that minority group members now enjoy enormously enhanced access to fine colleges, since it puts the top graduate from a poor, inner-city high school on the same competitive footing as his or her peer from a posh suburban school.

This system is no cure-all for the diversity dilemma. As some critics have pointed out, its efficacy as a method for enrolling Black and other minority group students actually depends on the persistence of de facto segregation in high schools—a phenomenon that in itself is a problem in search of a solution. And it can't be readily adopted at the graduate school level. But it is a useful interim tool and one that deserves to be more widely tested.

It's important for ADL to be in the forefront of supporting, publicizing, and promoting programs like these. After all, we have long supported the idea of increasing diversity, pluralism, and tolerance on campus and in the workplace, and we continue to favor these goals, so long as they are pursued without violating the principle of nondiscrimination on racial and ethnic lines.

I sense that, little by little, affirmative action is becoming less of a hot-button issue among African American leaders. Although very few now reject it outright, more and more are looking for alternatives. Nonetheless, the scars from this battle continue to mar the relationship between some Blacks and some Jews.

Support for the state of Israel has also become a flash point for relations between some in the Black and Jewish communities. Jews in the United States are loyal Americans who overwhelmingly place the interests of this country first when it comes to foreign policy. But as I've explained, the existence and security of the state of Israel are profoundly important to most Jews. Israel is both a repository of our cultural and religious heritage and, in time of need, a final refuge for imperiled Jews from around the world. Therefore, most Jews advocate a strong alliance between the Untied States and Israel. And since Israel is both the only democracy in the Middle East and the most stalwart ally of the United States in the region, most non-Jewish Americans feel the same way.

However, a significant proportion of African American political leaders have drifted away from this position in recent years. For example, in May 2001 twelve of the thirty Black congressional representatives voted against resolutions in support of Israel's antiterrorism campaign, a stunningly high percentage when you consider that the resolutions passed overwhelmingly. Cynthia McKinney, a five-term congresswoman from Georgia, and Earl Hilliard, a representative from Alabama, were two of these. Both had histories of sharply pro-Palestinian, anti-Israeli positions on the Middle East. Influenced by these positions as well as by their vote on the terror resolution, many Jews from around the United States opposed their reelection in 2002. Both

McKinney and Hilliard were defeated in the Democratic party primaries—in each case by another, more mainstream, African American candidate.

McKinney's defeat became a subject of national controversy when some Black spokespeople took umbrage at the fact that Jews from outside Georgia got involved in the election. Jesse Jackson even called it a crisis in the Black-Jewish relationship. The implication was that Jews who opposed McKinney did so because of her color, not because of legitimate disagreement over policy.

I think this perception is based on a set of misunderstandings. For one thing, most of those who opposed McKinney endorsed another Black candidate in the race, which should make it obvious that no anti-Black sentiment could have been involved. For another, Jewish activists have never been shy about criticizing any politician, white or Black, whose positions they disagree with. In any case, by most accounts, the role of Jews and other outside participants in the defeat of McKinney and Hilliard was a minor one. In Georgia, for example, the local system of crossover voting permits Republicans to vote in the Democratic primary—and it appears that many Republicans, angered by McKinney's criticisms of President Bush, did just that, playing a major role in her defeat.

I don't accuse McKinney or Hilliard of being anti-Semitic. I differ with them on American support of Israel, and like any other citizen of this country I have the right to express and act on my opinion. But the controversy over the role of Jews in opposing the reelection of McKinney and Hilliard illustrates how sensitive relations between Blacks and Jews have become.

AL SHARPTON: FAST AND LOOSE WITH HATRED

Currently mounting a campaign for the Democratic presidential nomination, the Reverend Al Sharpton encapsulates the troubled state of Black-Jewish relations in America today. I don't brand

Sharpton an anti-Semite. But in the style of all too many politicians, he has flirted with anti-Semitism in his checkered career as an activist and community organizer, drawing support from anti-Semites in the Black community and encouraging the spread of anti-Semitic beliefs and attitudes. Sharpton's rise to prominence is a disturbing emblem of the apparent mainstreaming of anti-Semitism among many Black Americans and a depressing contrast to the heady days when the battle for civil rights was led by men of the caliber of Dr. Martin Luther King Jr.

Sharpton's story is nothing if not colorful. He won fame at the age of four as a prodigy on the Pentecostal preaching circuit. At ten he was touring with the beloved gospel singer Mahalia Jackson. At fifteen he was named by Jesse Jackson the youth director of Operation Breadbasket, an organization that pressed for hiring of Blacks by corporations. During the 1970s, while continuing his political activism through his own National Youth Movement, he served as touring manager for soul singer James Brown.

Beginning in the 1980s, Sharpton became an increasingly visible presence on the political scene. It's hard to think of any controversy important to Black Americans over the past two decades in which Sharpton has played no role. He organized demonstrations in 1986 after a Black man named Michael Griffiths was chased onto a highway by a crowd of whites in the Howard Beach neighborhood of Queens, New York, and killed by a passing vehicle. He led protests against police brutality in the torture of Abner Louima (1997) and the killing of Amadou Diallou (1999). He has also sought to focus attention on some issues that most African American leaders shun, such as the persistence of slavery in Sudan.

For these exploits, Sharpton has received more than his share of attention, good and bad. In January 1991 he was stabbed in the chest by a drunken white man while preparing to lead a protest march in Brooklyn. He served a heavily publicized forty-five-day jail term in 1993 for his role in a march that shut down the Brooklyn Bridge. And in 1997 he won fully 32 percent of the city-wide vote in New York's Democratic mayoral primary. With his current presidential campaign, Sharpton seeks to solidify his claim to be the nation's most powerful Black leader.

Unfortunately, Sharpton's activism has included a series of notorious misadventures, some of which have helped produce real discord and anguish among both white and Black New Yorkers. Three incidents head the litany.

In 1988 Sharpton served as an adviser to Tawana Brawley, a Black teenager who claimed she had been abducted and raped by a gang of whites in her upstate New York hometown. Describing the alleged incident as an example of white-on-Black violence countenanced by the power structure, Sharpton continued to support Brawley even as her story morphed and grew, ultimately going so far as to accuse state prosecutor Steven Pagones of participating in the crime. A grand jury found no evidence to support Brawley's claims, and in 1998 Pagones won a $65,000 defamation claim against Sharpton.

In 1991 Sharpton helped fan the flames of anger in Crown Heights, calling for the Black community to rise in protest after young Gavin Cato was killed in a car accident involving a Hasidic driver. As we've seen, the resulting three days of violence culminated in the stabbing death of student Yankel Rosenbaum.

And in 1996 Sharpton was the most prominent leader of protests in Harlem against Freddy's Fashion Mart, a Jewish-owned clothing store that local leaders considered a symbol of Black economic disenfranchisement. Sharpton referred to the store's owner as "a white interloper," and some in the crowd of protesters were heard to shout, "Burn it down!" Sure enough, a deranged man later set the store on fire, killing himself and seven other people—ironically, none of them Jewish.

In his quest for mainstream respectability, Sharpton has tried to distance himself from these tragic events. He stresses that his claims about Pagones preceded the grand jury finding that Tawana Brawley's story was false. In regard to Crown Heights and Freddy's, he emphasizes his personal commitment to nonviolent protest and tries to downplay the bigotry in his remarks: "I called him [Freddy] an interloper. I shouldn't have referred to his race."

Are these retractions sufficient to absolve Sharpton of having at least flirted with antiwhite and anti-Semitic hatred? Not for me. A man who aspires to head a national political ticket and, even more important, to claim the mantle of moral leadership once worn by Dr. King needs to do a much better job of confronting and expiating his own sins.

Both Sharpton and the Democratic Party are well aware of his bad reputation among many whites, especially Jews, and they are rightly worried about the impact Sharpton's prominence will have on the party's fortunes. In 2000, when Hillary Clinton was running for the United States Senate in New York, I got a call from some of her Jewish supporters, asking what would it take for me to "make peace with Al Sharpton." Their goal: to reduce tensions between the Black and Jewish communities in New York and thereby make it easier for Hillary to carry the all-important Jewish vote in the election.

My first reaction was to protest, "I'm not at war with Al Sharpton." That's true, of course. I've criticized Sharpton publicly when I felt he was doing something wrong. But I never mounted any kind of ongoing campaign against him, nor would I do so.

But I knew what the Hillary supporters meant. They were seeking a kind of Abe Foxman seal of approval for Sharpton—to have me lay to rest, on behalf of ADL, the imputation of anti-Semitism that hung over him and, by extension, his Democratic Party supporters.

Naturally, I couldn't speak for the entire Jewish community. But as an informed observer, I could give an opinion as to how thoughtful Jews were likely to react to Sharpton's support for Hillary. So I offered this diagnosis: "If Al Sharpton wants credibility with the Jewish community, there are three things he needs to do. First, he has to apologize for his statements during the Crown Heights violence. Second, he has to apologize for his role in the burning of Freddy's in Harlem. And third, he has to admit his wrongdoing in the Tawana Brawley case."

The intermediaries conferred with Sharpton and called me back. "Al will agree to the first two items," they told me, "but not the third. He says that Tawana Brawley is not a Jewish issue."

"That's true," I said. (The defamed prosecutor Steven Pagones was not Jewish, and anti-Semitism played no direct role in the Brawley case.) "But it was an example of Black racism. For me and for most Jews, bigotry in any form is dangerous and abhorrent."

In the end, I was unable to give Sharpton the blessing he sought. But my advice to Sharpton remains the same. He'll follow it if he wants to play a truly constructive role in rebuilding trust between the Jewish and Black communities in America.

Sharpton is an interesting figure who may yet play a significant role on the national political scene. But I'm more concerned about what he symbolizes. The prominence being accorded a man with a history of flirtation with antiwhite and anti-Semitic bigotry is very disturbing. The existence of a seemingly broad base of support for Sharpton in the Black community troubles me. It's a worrisome symptom of latent anti-Semitism among a significant number of African Americans.

PLAYING WITH FIRE: ANTI-SEMITISM IN BLACK CULTURE

Another such symptom is the growing tolerance for racist and anti-Semitic stereotypes, epithets, and doctrines in Black culture, whether on college campuses or in the world of entertainment.

Since the 1980s anti-Semitism has been a growing campus force among African American students. The Nation of Islam, which I'll discuss in detail in a moment, is an important player in this arena. Others include a number of Black academics who invoke academic freedom as justification for espousing racist and anti-Semitic views in the classroom and in outside lectures.

Leonard Jeffries, the former head of the Black Studies department at the City College of CUNY, and a professor there since 1972, has long used antiwhite and anti-Jewish rhetoric in his teaching and writing. He became notorious in 1990 and 1991 for a series of lectures in which he asserted that "rich Jews" controlled the Black slave trade and that Hollywood was run by a Jewish-dominated conspiracy that systematically denigrated Blacks.

When CUNY's board of trustees voted to remove Jeffries from his job as head of the department, he sued on free speech grounds. After a three-year legal battle, Jeffries finally lost the post in June 1995. However, he remained a tenured professor at City College. And his hateful views apparently had not changed. Jeffries was a featured speaker at the viciously anti-Semitic and racist Black Holocaust Nationhood Conference held in Washington, D.C. in October 1995, the weekend before the Nation of Islam's Million Man March. To this day he continues to spread the same perverted teachings through lectures on campuses around the country.

Another well-known Black educator who uses his position to retail anti-Semitic hatred is Anthony Martin, a tenured history professor in the Africana Studies department of Wellesley College. Martin became the subject of controversy when it was learned that he had assigned as a primary textbook for a survey course on African American history *The Secret Relationship Between Blacks and Jews,* an anonymous diatribe that purports to show that Jews dominated the slave trade. In response to criticism, Martin self-published his own book, titled *The Jewish Onslaught,* which describes a "conspiracy" against him and calls Blacks who disagree with him "handkerchief heads" and "unthinking Negro stooges."

Are there also racists and demagogues among white educators? Undoubtedly, and I've often denounced them. But Black racists such as Martin and Jeffries have managed to win more prominent, influential, and prestigious academic posts than their white counterparts and, with those posts, the veneer of respectability that makes these educators dangerous sources of hatred for impressionable college students in search of their own cultural identity.

A similar kind of tolerance has been extended to anti-Semitism and racism in some quarters of the Black entertainment industry—in particular rap or hip-hop music.

It may be unnecessary to discuss the violent, hateful nature of much hip-hop music. Ever since the emergence of so-called gangsta rap in the mid-90s, the prevalence of misogynistic, homophobic, and racist language in rap has been widely discussed and debated. Still, I find the popularity of this style of music among audiences both Black and white particularly depressing. The Recording Industry Association of America reports that rap CDs have accounted for close to 10 percent of all record sales since 1999 (though rap sales reportedly tailed off slightly during 2002).

Not all rap artists are Black, of course. (Today's most popular rap singer is the white Eminem.) And not all rap songs espouse violence or hatred. But too many popular Black hip-hop artists have released records with veiled or open anti-Semitic messages, like these lines evoking the Holocaust from the song "Swindlers Lust" by the popular and influential group Public Enemy:

> More dollars, more cents for the big six
> Another million claiming they innocence
> Is it any wonder black folks going under

Or the same group's "Welcome to the Terrordome," which dusts off the old accusation of deicide against the Jews:

> *Crucifixion ain't no fiction*
> *So called chosen frozen*
> *Apology made to who ever pleases*
> *Still they got me like Jesus*

Incidentally, the man who wrote these lyrics, Public Enemy leader Chuck D., born Carleton Ridenhour, recently launched a new band as a side project. Its name? Concentration Camp.

It would be unfair to hold an entire community responsible for the horrific messages embedded in one brand of music. (To cite a parallel, whites as a group aren't responsible for the hate music peddled by the National Alliance's Resistance Records.) But the fact that hip-hop containing racist, homophobic, and violent language is widely regarded as the mainstream music of today's Black youth—as well as of many whites—is a deeply troubling cultural symptom.

Even some Black cultural icons with much more benign images have been known to flirt with anti-Semitism.

Take Michael Jackson. Once known as the King of Pop, Jackson was for decades one of the most popular and talented singers, songwriters, and performers in the world, a crossover star with a huge following among members of every race and ethnicity. So many of us were dismayed when, in the spring of 1995, Jackson's album *HIStory* was released with the song, "They Don't Care About Us," including the lyrics, "Jew me, Sue me . . . Kick me, Kike me."

Naturally, an uproar ensued. At first Jackson defended the song, claiming that it was intended as a protest against all forms of prejudice. Maybe so. But the "protest" was ambiguous at best. Jackson's use of the hateful word *Kike* (as well as his use of *Jew* as a verb) could easily encourage the same kind of language—and the attitudes behind it—among his millions of worldwide fans. I was amazed at Jackson's ignorance and foolishness but even more at the failure of any of the scores of people involved in producing and marketing the record to raise a protest against the use of such epithets. (That failure illustrates the truth that, *pace* the assertions of anti-Semitic conspiracy theorists, there is no Jewish cabal dictating to the entertainment industry—or if there is, it is a remarkably ineffective one.)

Scalded by widespread criticism, including a public rebuke issued by ADL, Jackson wrote a letter of apology:

> My sole intention with the song "They Don't Care About Us" was to use language to demonstrate the ugliness of racism, anti-Semitism, and stereotyping. I had hoped that my lyrics would target the bigots, not the victims of bigotry.
>
> Because of the unforeseen reaction to specific words in my song by many people around the world, I have chosen to rerecord it, deleting the words found offensive. In doing so, I acknowledge that I seriously offended some people which was never my intention and for that I am deeply sorry. I have come to understand over the past days that these words are considered anti-Semitic.

Of course, I accepted Jackson's apology. But in assuming that he would act on his words, I was mistaken. Eight months later, in February 1996, we learned that the epithets had been reinstated for the album's video release and that the words "Jew me, Sue me . . . Kick me, Kike me" had been sung on a nationally syndicated television program.

Michael Jackson likes to claim that he is a spokesperson against hatred and bigotry. But after this episode, he has no credibility in my eyes.

MINISTER OF HATE

I'm troubled by Black entertainers who flirt with anti-Semitism or use it in their music for its shock value. But even more frightening are supposed African American spokespeople who are out-and-out anti-Semites—truly bigoted people who spout hatred at every opportunity. Today the most dangerous anti-Semitic voice among American Blacks is that of Louis Farrakhan, the influential leader of the Nation of Islam, the religious sect also known as the Black Muslims.

Founded in 1930 and led for over four decades by Elijah Muhammad, a Black Southerner originally known as Elijah Poole, the Nation of Islam (NOI) is a homegrown American religion, as are the Mormons, the Christian Scientists, and Scientology. The teachings and practices of the NOI blend a strong message of self-help, racial pride, personal discipline, and moral conservatism with a weird cosmology: Blacks are supposed to be descended from Shabbaz, a tribe that came from the moon 66 trillion years ago, while white people are considered devils who were concocted in a laboratory by the evil scientist Yakub.

Although the NOI claims an affinity with the worldwide Muslim faith as founded in the seventh century by the prophet Mohammed, most Muslim leaders and scholars consider it at most a loosely connected offshoot. Today the NOI runs some fifty mosques and a collection of businesses that includes a newspaper, a toiletries company, a Chicago restaurant, a Georgia produce farm, and a private securities firm. NOI also runs twenty-five student organizations on college campuses, and spokesmen from NOI are frequent guest lecturers at Black student unions around the United States.

Although boxer Muhammad Ali is undoubtedly the most famous adherent of the NOI, the life of political activist Malcolm X probably illustrates most clearly the strange mixture of qualities found in Black Muslim culture and teachings.

Undoubtedly the NOI played an enormously positive role in Malcolm's early life. Born Malcolm Little, he was a small-time hoodlum, pimp, and drug dealer when he was introduced in prison to the teachings of Elijah Muhammad. Galvanized by what he learned, Malcolm reformed his life, gave up crime, and transformed himself into an effective preacher, street-corner orator, and civil rights agitator. His uncompromising advocacy of Black community power and armed self-defense frightened many in mainstream America.

Over time Malcolm X began to abandon or modify the racist tenets of the NOI. In particular, a *hajj*, or religious pilgrimage, that Malcolm undertook to the holy city of Mecca in Saudi Arabia in 1964 seems to have to moved him to a sense of kinship with other Muslims of every color. As a result Malcolm returned to America with a new willingness to explore the possibilities of racial peace and cooperation between whites and Blacks. He broke with the NOI and was laying the foundation for his own Islamic religious and political organization when he was assassinated in 1965, apparently by a group of NOI loyalists.

It's tempting to speculate about how the Nation of Islam might have been transformed if a more moderate Malcolm X had lived to carry out his plans. Instead, a new generation of Black Muslim leaders came to power that had no intention of abandoning the racial bigotry that has always marred the record of the NOI.

Since 1975 the NOI has been led by Louis Farrakhan. Born Louis E. Walcott in Roxbury, Massachusetts, in 1933, Farrakhan was a talented violinist and calypso singer when young. At the age of twenty-two he attended an NOI convention and soon joined the sect, abandoning his performing career for a life of preaching, teaching, and proselytizing. Since becoming head of the NOI, Farrakhan has pushed the organization's agenda on several fronts. He has expanded the NOI's network of mosques and study groups in American cities and opened several NOI-owned businesses designed to exemplify the Muslim message of Black economic self-development. He has also traveled widely on behalf of the NOI, trying to link the Black Muslims with traditional Islamic communities in the Middle East, Africa, and Asia.

Farrakhan has also worked to position himself as a leading spokesperson for the African American community, speaking out on college campuses and in other public forums about practically every issue of interest and concern to Black Americans. Unfortunately, these public pronouncements have continually been tinged with racist and specifically anti-Semitic attitudes. Here are some of the classically anti-Semitic positions Farrakhan has taken:

- Farrakhan claims that Jews largely control the American media and government. Thus whenever Black Americans suffer at the hands of the white establishment, Jews are primarily to blame.

- He claims that a worldwide conspiracy led by Jews controls international finance and business. Farrakhan likes to point out that the FBI, the Internal Revenue Service, the Federal Reserve, and the Anti-Defamation League were all founded in 1913, which he considers not a coincidence but evidence of this conspiracy.

- He claims that the Black history of exploitation and oppression is largely the fault of the Jews. Specifically, he claims that Jews dominated the worldwide slave trade. (As historical fact, this is ludicrous. The role of Jews in the slave trade was tiny compared to that of Christians—or of Muslims, for that matter.)

- He claims that Jewish doctors deliberately infected Black Americans with the AIDS virus.

· He denies or minimizes the reality of the Holocaust, claiming that the figure of six million Jews murdered by the Nazi regime is exaggerated.

Farrakhan's anti-Semitic positions are well documented since he has espoused them not just in one or two interviews or speeches but repeatedly.

Also well documented is his flirtation with terrorist leaders around the world, particularly among the Arab nations of the Middle East. During his twenty-seven-day world tour on behalf of the NOI in January 1996, Farrakhan visited Libya, Sudan, Iraq, and Iran. He met with and publicly praised such tyrannical strongmen as Libya's Moammar Qaddafi, President Omar Hassan al-Bashir of the Sudan, and Iraq's Saddam Hussein, all well-known supporters of anti-Israeli and anti-Western terrorism. Qaddafi reportedly promised the NOI a one-billion-dollar donation, but United States sanctions against the Libyan regime prevented it from materializing.

Soon after his return from this tour, Farrakhan gave a speech at the District Council 33 Union Hall in Philadelphia, Pennsylvania, in which he described the notorious Hezbollah terror organization in this way:

They call them terrorists, I call them freedom fighters. . . . No one asks why they would do such a thing. What has driven them to this point? That's what the UN, the US, and Europe doesn't want to deal with because the Zionists have control in England, in Europe, in the United States, and around the world.

The message of hatred comes not only from Farrakhan himself but also from various arms of the Nation of Islam, all of which Farrakhan controls. Consider, for example, his lieutenants who speak on campuses and other sites, the most notorious being the late Khalid Abdul Muhammad, formerly information minister for the NOI. His most well known speech (at New Jersey's Kean College in November 1993) was filled with bizarre and ugly anti-Semitic slurs:

Who are the slum lords in the Black community? The so-called Jew. . . . Who is it sucking our blood in the Black community? A white impostor Arab and a white impostor Jew. Right in the Black community, sucking our blood on a daily and consistent basis. . . . Brother, I don't care who sits in the seat at the White House. You can believe that the Jews control that seat that they sit in from behind the scenes. They control the finance, and not only that, they influence the policy-making. . . . These people have had a secret relationship with us. They have our entertainers in their hip pocket. In the palm of their hand, I should say. They have our athletes in the palm of their hand.

The same speech included weird personal attacks on gays, Catholics, and a host of individuals ranging from Pope John Paul II to movie-maker Spike Lee and actress Elizabeth Taylor. Clearly Khalid Muhammad was an equal-opportunity hate-monger.

Farrakhan's anti-Semitism also appears through the distribution of vicious anti-Semitic books, such as the notorious forgery *The Protocols of the Elders of Zion,* at many Nation of Islam functions, along with the NOI publication *The Secret Relationship of Blacks and Jews,* which claims that Jews largely controlled the African slave trade. It surfaces on a regular basis in the organization's newspaper, the *Final Call,* which is full of anti-Semitic blather.

For a number years Farrakhan has tried to project a more moderate image when appearing in mainstream circles. He couches his anti-Semitic slurs in vague language, even trying to make his attacks on Jews sound like compliments about their intelligence, business acumen, and political savvy. He even tried to distance himself from his own spokesman, Khalid Muhammad, after the Kean College speech was widely denounced in the media—only to issue a personal tribute to him after his death in February 2001.

To some extent Farrakhan's attempt to spin his anti-Semitism has been successful. He has been given a forum on such mainstream media outlets as *Good Morning America, Meet the Press,* and *20/20,* where he has been treated with greater respect and deference than his stature would seem to justify. And he has often been treated as an important legitimate spokesman for the Black community—for example, on September 2, 1995, when he was invited by the Congressional Black Caucus to participate in a panel with recognized Black leaders.

Perhaps Farrakhan's most ambitious attempt to appear as a mainstream Black leader was his Million Man March in Washington, D.C., on October 16, 1995. Its stated goal was to foster a spirit of unity and self-improvement among Black American males. An estimated seven hundred thousand participated, and the media coverage of the event was enormous and largely positive.

Only 5 percent of the participants polled on the mall said that they were marching because of Farrakhan. But over 30 percent expressed agreement with his anti-Semitic views. It appears that the majority of Black Americans continue to reject the hateful messages Farrakhan and the NOI try to spread and that the attempts of Farrakhan and his followers to co-opt such worthwhile causes as Black freedom and economic development have so far failed.

Nonetheless, Farrakhan and other NOI spokespeople didn't hesitate to use the Million Man March as an occasion to spout many of the same anti-Semitic slurs they have spread for decades. And unfortunately, a significant minority among their audience has no problem with these attitudes. One of the most disturbing things about a Farrakhan speech is the fact that the noise level of the crowds who are listening is an accurate barometer of the amount of hatred in the statements. When Farrakhan begins talking about "whitey," the noise level increases. And when he starts talking about Jews, the crowd goes ballistic.

It would be a mistake to believe that Farrakhan's anti-Semitism was a youthful indiscretion that he has disavowed or outgrown. Here is a sampling of public utterances by Farrakhan that date from the midnineties—*after* the NOI's Million Man March and other efforts to attain mainstream respectability:

- From a speech in February 1998: "Of course, they [the Jews] have a very small number of people but they are the most powerful in the world, they have the power to do good and they have the power to do evil. . . . Now what do the Jews do best? Well, they have been the best in finance that the world has ever known."

- From an interview in October 1998: "They [the Jews] are the greatest controllers of Black minds, Black intelligence. They write the scripts—the foolish scripts on television that our people portray. They are the movie moguls that feature us in these silly, degrading, degenerate roles."

- From a speech in August 2000: "Is the Federal Reserve owned by the government?" (Audience: "No.") "Who owns the Federal Reserve?" ("Jews.") "The same year they set up the IRS, they set up the FBI. And the same year they set up the Anti-Defamation League of B'nai B'rith. . . . It could be a coincidence. . . . [I want] to see black intellectuals free. I want to see them not controlled by members of the Jewish community."

- From a speech in February 2003: "I don't hate Jews. I honor and respect those who try to live according to the teachings of the Torah, but you can't criticize Jewish people. If you criticize

them you are anti-Semitic. . . . The Bible says, Revelations,
those who say they are Jews and are not, I will make them of
the synagogue of Satan. I don't hear you preaching that full
Gospel. You are afraid of consequences."

Some well-meaning people have tried to get me and other
Jewish leaders to meet with Farrakhan as part of our ongoing
efforts to promote dialogue and cooperation between the Black
and Jewish communities. Some, misled by the sanitized versions
of his beliefs presented on network television, argue that he really
isn't so bad. Others maintain that if former enemies like Israeli
prime minister Rabin and Palestinian leader Yasser Arafat could
meet, why not Minister Farrakhan and the Jews? Still others
assert that, simply because Minister Farrakhan is so dangerous
and so well known in the Black community, Jews cannot afford *not*
to meet with him.

I'm certainly a believer in open dialogue, even between people
with views that seem to be hopelessly divergent. But I'm convinced that meeting with Farrakhan would damage the interests
of the Jewish community.

First, for those who would minimize Farrakhan's bigotry, it is
important to recognize that our concern is not about an occa-
sional statement by him. It is rather about a systematic record
spanning over twenty years of bigotry, scapegoating, conspiracy
theories, anti-Semitism, Catholic bashing, and homophobia.

As to the Rabin-Arafat analogy, what this ignores is the absence of the very thing that Farrakhan described as the theme of
the Million Man March—atonement.

Whatever his current shortcomings, Chairman Arafat, after
many years of murderous terrorism in service to the goal of
destroying Israel, finally committed himself in 1993 to give it up:
no more terror, nor more rejection of Israel, no more war. Then
and only then did Israeli leaders agree to meet with him. Farrakhan

has shown no such repentance and no willingness to stop spewing his hatred.

As to the notion that Jewish groups can't afford to ignore Farrakhan, this is a recipe for capitulation and disaster. The fact that Minister Farrakhan has gained legitimacy in the eyes of some people without changing his message of hate makes it more necessary than ever for good people to stand up against him.

A meeting with Jewish leaders would signal the public that Farrakhan belongs on center stage. We shouldn't present him with the gift of legitimacy he has done nothing to deserve—nor should we undermine the efforts of responsible Black leaders who condemn Farrakhan's bigotry.

Rebuilding the Partnership

Farrakhan and the NOI aren't the only well-known Black Americans who have used their prominence to promote anti-Semitism. Another is Wilbert Tatum, the editor of the *New York Amsterdam News*. Although he happens to be married to a Jew, he has used the forum of his paper as a venue for spreading anti-Semitic and near-anti-Semitic attacks, both in the editorial and opinion columns of the paper and in the slanting of stories on the news pages.

In particular, the *Amsterdam News*'s coverage of the 1991 Crown Heights crisis seized on the tragedy as an opportunity for attacking the Jewish community. It also unthinkingly defended every action taken by members of the Black community—even seeming to minimize the guilt of the cold-blooded killers of Yankel Rosenbaum. In so doing, the paper played into traditional anti-Semitic images of Jews as self-centered, power-hungry, and heartless, thereby heightening rather than calming tensions throughout the city. More recently, Tatum attributed the 2000 vice-presidential candidacy of Joseph Lieberman to the power of "Jewish money," a classic anti-Semitic slander.

I worry about Tatum's editorial stance in such issues because of the damage it does to the already fragile state of relations between our two communities. There is only one *Amsterdam News*. Although it certainly doesn't speak for all in Black America, it has an influence and a voice that resonate across the country. I am dismayed that this powerful voice is serving in too many instances to inflame rather than to quell interracial conflict, and, in some cases, to feed fuel to the fires of anti-Semitism.

Thankfully, American Jews continue to have many good friends in the Black community. ADL is working closely with groups ranging from the National Urban League and the NAACP to the National Conference for Community and Justice (formerly the National Conference of Christians and Jews) to rebuild the multiethnic alliance that launched the civil rights movement of the 1960s.

A number of individual Black leaders are good friends of ADL and the Jewish community. Many, such as Representative Alcee Hastings, stood up in our defense in the McKinney case. I've already mentioned Hugh Price, another steady supporter of ours, and Secretary of State Colin Powell, a man of integrity whom all Americans admire. General Powell and I both served in the Reserve Officers Training Corps (ROTC) when we were students at the same time at New York's City College. (He went a little further in the military than I did.) Today he is an important connection for ADL in our efforts to maintain a strong and even-handed American approach to conflict in the Middle East.

Then there was the late Mickey Leland, who went through a fascinating political odyssey. He grew up as a member of the radical Black Panther party. Later he moved toward the political center and was elected to Congress from Houston, Texas. In the 1990s, when ADL was working to help liberate the Jews of Ethiopia, Leland helped us gain access to Colonel Megistu Haile

Mariam, that country's Marxist dictator. Soon the Ethiopian Jews were on their way to a new life in Israel.

Ironically and tragically, Leland's involvement in the Ethiopian case indirectly cost him his life. After the Jews were liberated, he got deeply involved in the welfare of the Ethiopian people and died in a plane crash in 1989 while bringing famine relief to the country.

In truth, the relationship between Blacks and Jews has always been a complicated one, with strong elements of love, admiration, resentment, and envy on both sides. To some extent, we argue and fight precisely because we care about one another so much. In fact, if this weren't so, I and many others wouldn't spend so much time analyzing and discussing the Black-Jewish relationship. Think about it: no one talks about how most other ethnic groups get along. There are no articles written or seminars held to deal with the Irish-Jewish or the Irish-Black relationship. No one debates about how Italian Americans or German Americans are getting along with Jews or Blacks. Those relationships are simply not fraught the way the Black-Jewish relationship is.

The needs and problems of America's Black community are enormous, and they've been shamefully neglected by most political leaders for at least the last three decades. Like all people of goodwill, I sympathize with the frustrations and anger many African Americans feel. In fact, if I believed that blaming the Jews would save the life of one kid in Harlem, I would be inclined to say to my fellow Jews, "Let's tolerate the attacks—for the sake of our Black brothers and sisters."

But in fact anti-Semitism does no one any good. If anything, it retards Black progress by distracting us all from the real political, social, and economic steps we should be taking as a nation. The sooner we can make Black anti-Semitism a topic for the history books rather than the front page of today's newspaper, the better off we all will be.

CHAPTER 7

From Hatred to *Jihad:*
Anti-Semitism in the Muslim World

FOR MANY YEARS ADL AND OTHERS at the forefront in the war against bigotry and hatred treated anti-Semitism in the Arab world as a marginal issue. When we looked at the Middle East, we focused on the Arab-Israeli conflict, seeking ways to launch an effective peace process, and we assumed that improved relations would follow a negotiated peace. Perhaps thinking wishfully, we attributed anti-Jewish bigotry among Arabs to the Israeli-Palestinian conflict rather than to any deep-seated prejudice. Many claimed that Jews living under Islam through the centuries had experienced far more tolerance than Jews in Christian Europe. And those who wanted to focus on Arab anti-Semitism were often dismissed as obstructionists who were eager to avoid the concessions that a true peace process would require.

Now the events of the past two years—the outbreak of the new *intifada*, the September 11 attacks, the unraveling of the Oslo peace process, and the current frightening upsurge of Arab anti-Semitism—have forced us to take a fresh look at the connection between anti-Semitism and the Arab-Israeli conflict.

The fact is that virulent anti-Semitism is widespread throughout the Arab Middle East. (Note that the common Western practice of equating Muslims with Arabs is strictly inaccurate. Fully 80 percent of the worldwide Muslim population today is non-Arabic, with Indonesia being the world's largest Muslim nation. Nonetheless, in this chapter, we will use Islam and the Arab world as overlapping terms, since our focus here is on the Muslim nations of the Middle East.) Anti-Semitism is tolerated or openly endorsed by Arab governments, disseminated by the media, taught in schools and universities, and preached in mosques. No segment of society is free of its taint.

Not every Arab is an active hater of Jews. Some may reject anti-Semitism entirely. But to the extent that we are able to speak of an attitude or ideology permeating a society, informing the beliefs of the masses, the debates of the intelligentsia, and the decisions of the leaders, it's clear that the Middle East is permeated with anti-Semitism.

As always, I want to distinguish anti-Semitism from opposition to Israeli policies and actions. Principled, fair criticism of Israel and Israeli leaders is always permissible. But when *Zionist* becomes a curse word and Zionists can be blamed, as if by reflex, for the terror attacks of September 11; when Israelis are caricatured using imagery drawn from Nazi propaganda; when Israel's prime minister is depicted as an evil puppet master driving the world down a path toward Jewish domination—and when all of these bizarre distortions of reality are repeated in the mainstream Arab press day after day—then we have clearly moved beyond honest opposition to policies and actions and entered the realm of pure anti-Semitism.

The implications of Arab anti-Semitism for Mideast peace and for the safety of world Jewry are deeply disturbing. In today's wired world, anti-Semitic pronouncements that originate in mosques or on street corners in Egypt, Saudi Arabia, and Syria are quickly disseminated globally, helping to fuel hatred of Jews among over a billion Muslims from Cincinnati to Singapore, from Buenos Aires to Budapest. These Muslim communities constitute a rapidly growing political force in dozens of countries, eager to push governments toward anti-Jewish, anti-tolerance policies. Some of these governments have access to weapons of mass destruction. Most frightening of all, Arab hate-mongering is promoting the rise of yet another generation of terrorists and suicide bombers among Muslim youth. The only possible outcome is still more needless violence and death—not only for Jews, but for all peoples in the Middle East and around the world.

TRADITIONAL MUSLIM ATTITUDES TOWARD THE JEWS

How did it come to pass that Jews have come to be considered the eternal enemies of Islam and Allah? Like Christian anti-Semitism, Muslim anti-Semitism has a long and complex history.

The demonization of Jews was not a traditional component of Islam. Islam as a religion has viewed both Jews and Christians as "peoples of the Book," bearers of partial versions of divine truth as revealed in the Hebrew and Christian scriptures. However, in the traditional Muslim view, Judaism and Christianity were both distorted by human frailty and ultimately superseded by Islam, which is the perfect expression of the one true religion. Therefore, Jews and Christians were permitted to live in Muslim lands as tolerated minorities (*dhimmis*), free to practice their religions but subject to the humiliations of second-class status.

Jews in particular had to live with the legacy of Muhammad's historical interactions with their coreligionists from Medina; the ire he felt at their opposition to his expanding influence, recorded in the Qur'an (the Islamic scriptures), was followed by his triumph over them and their subjugation to his word. This hostility and triumphalism set the tone for Islam's subsequent attitude toward the Jews. As descendants of those who distorted God's truth and opposed his Prophet, Jews would rightly be humbled before Muslims.

Of course, Muslim attitudes toward and treatment of the Jews varied from time to time and place to place. Not all Muslim governments were equally eager to impose strict interpretations of the *dhimmi* paradigm on their Jewish subjects. Many Muslim societies, especially during the High Middle Ages (tenth to twelfth centuries), exhibited a rare tolerance for Jews and other minorities. When Spain was largely under Muslim rule, for example, Jews enjoyed what some consider a golden age of freedom and prosperity. And after 1492, when the Jews were expelled from Christian Spain, it was the Islamic Ottoman Empire that accepted many of them as refugees.

However, even during this era of relative freedom, the toleration of Jews was limited. For example, there were rules restricting the number and location of synagogues, which could not be built in close proximity to mosques. And alongside those Muslims who treated their Jewish neighbors and associates with friendship and acceptance there were always others who stressed that the proper Islamic approach called for Jewish debasement. Thus an element of tension always existed between Muslims and Jews in the Middle East, not unlike the tension between Christians and Jews in Europe.

THE NEW DEMONIZATION

During the twentieth century Muslim attitudes toward Jews underwent a horrific transformation, driven by broad historical forces, some of which have little or nothing to do with Jews themselves.

Nazi-style anti-Semitism traces its history in the Middle East back to 1937, when Nazi leaders conducted propaganda campaigns in the region. The mufti of Jerusalem during World War II, Hajj Amin al-Husayni, tried to establish an alliance between Nazi Germany, fascist Italy, and Arab nationalists, for the ultimate purpose of conducting a Holy War of Islam against "international Jewry." Several Nazi-influenced political parties arose in the Middle East in the 1930s and 1940s, some of which went on to play important roles in shaping the leadership of Arab nations in the post–World War II period. Egypt, Syria, and Iran are widely believed to have harbored Nazi war criminals, though they do not admit doing so. Hitler's hate-filled autobiography, *Mein Kampf,* has been published and republished in Arabic since 1963.

Nazi influence did not make anti-Semitism into a dominant force in the Middle East. However, during the past forty years, hatred of Jews has risen throughout the region in response to the emergence of a largely new Islamic theology in which anti-Semitism is treated as divinely ordained. And in most of the Arab world, this new theology dominates public discourse while voices of tolerance are vanishingly faint.

This theology of hate grew out of burgeoning fundamentalist movements within Islam, themselves spurred by widespread disillusionment and frustration over centuries of poverty and Western imperialism in the Middle East. It was expressed most famously in the proceedings of the Fourth Conference of the Academy of Islamic Research (1968), usually known simply as the Al-Azhar conference. Held at the most prestigious mosque and university serving the dominant Sunni branch of Islam (sometimes called the Harvard of the Islamic world), this conference in Cairo brought together Islamic theologians and professors from twenty-four countries in a forum that presented the quasi-official positions of most Islamic peoples.

In the speeches and writings that emanated from this conference, the alleged Jewish distortion of God's initial revelation to them and the Jewish opposition to Muhammad were removed from their historical contexts and transformed into symbols of an essential evil in Jewish nature.

Theologian Muhammad Azzah Darwaza wrote at the Al-Azhar conference:

[The Qur'an reveals that the Jews of ancient times] coated what was right with what was wrong. The Jews were also stubborn in telling lies and contradicting the truth. . . . They told lies about Allah. . . . They were notorious for covetousness, avarice, and bad manners. They were not ashamed of embracing polytheism or performing the rites of paganism. They sometimes praised the idols and were in collusion with idolaters against monotheists. They displaced the words of Allah and disfigured the laws of Heaven and God's advice. They were hard-hearted and sinful, they committed unlawful and forbidden crimes. . . . It is extremely astonishing to see that the Jews of today are exactly a typical picture of those mentioned in the Holy Qur'an and they have the same bad manners and qualities of their forefathers although their environment, surroundings, and positions are different from those of their ancestors. These bad manners and qualities of the Jews ascertain the Qur'anic statements about their deeply rooted instinct which they inherited from their fathers. All people feel this innate nature of the Jews everywhere and at all times.

Muhammad Sayyid Tantawi, a Muslim cleric who eventually became the mufti of Egypt and sheikh of the Al-Azhar University, described the Jews as follows:

Anyone who reads the Qur'an will clearly see that it attributes many negative moral qualities, ugly characteristics, and malicious methods to the Children of Israel. It describes them in terms of unbelief, rejecting the truth, selfishness and arrogance, cowardice and lying, obstinacy and deceit, disobedience and transgression, hardness of heart, deviance of character, competing in sin and aggression, and wrongfully consuming people's wealth. . . . [The Qur'an thus] firmly connects the morals and characteristics of those [Jews] who lived during the time of the Prophet [Muhammad] with the morals and characteristics of their first forefathers . . . to demonstrate that the sons' moral depravity, disobedience, and opposition to Islam constituted a legacy of the deviant character inherited by later generations from earlier ones. . . . [These characteristics] apply to them—as indeed we are witnessing—in all times and places, and the passage of time increases the deep-rootedness of these qualities in them. . . . The Qur'an mentioned these [characteristics of the Jews] to . . . warn believers against their evils and abominations.

Other Islamic traditions about Jews throughout the ages were altered and incorporated into a new narrative of Jewish malevolence toward Allah and to Muslims generally. Positive Islamic traditions about Jews were revised to reflected the new anti-Jewish attitude. For example, the story of Samaw'al, a Jew who was held by Arabs as the paradigm of fidelity for his willingness to allow his son to be killed rather than surrender items entrusted to his safekeeping, was reinterpreted in a negative light: Samaw'al's action demonstrated merely that he loved money more than his son's life.

Ultimately, Jews came to be described as the "eternal" enemies of Allah and of Islam, a satanic, diabolical force, locked in a lethal struggle with Islam. Theologian Sayyid Qutb, sometimes described as the father of Islamic fundamentalism, wrote, "The struggle between Islam and the Jews continues in force and will thus continue, because the Jews will be satisfied only with the destruction of this religion [of Islam]." In the same vein, Shaykh Abd-al-Halim Mahmud, the rector of al-Azhar University in Egypt, identified the Jews as Islam's worst enemies:

> As for those who struggle against the faithful [Muslims], they struggle against the elimination of oppression and enmity. They struggle in the way of Satan. Allah commands the Muslims to fight the friends of Satan wherever they may be found. And among Satan's friends—indeed, his best friends in our age—are the Jews.

One participant at the Al-Azhar conference described Jews as "hostile to all human values in this world," and another, the mufti of Tarsus in Syria, claimed that Jews "have always been a curse that spread among the nations and . . . sought to . . . extinguish all manifestations of civilization." Similarly the imam of the main mosque in Amman blamed the Jews for all the evil in the world:

> Jews are treacherous, ungrateful killers of their prophet. . . . Wherever they went they generated disaster. They stand behind all conspiracies and corruption in the world. God protect us from their evil!

The Al-Azhar conference of 1968 was not the first time anti-Semitic beliefs had been expressed in the name of Islam. But it represented a watershed in Muslim-Jewish relations because it was the first time that so many Islamic teachers and theologians had chosen to make the Jews a unique focus of enmity.

What Drives Arab Anti-Semitism?

What led to the deterioration of the traditional Islamic view of Jews into this hateful bigotry? Many assume that the answer is Israel—that Arab resentment over the establishment of a Jewish state in the midst of a predominantly Islamic Middle East is the driving force behind Muslim anti-Semitism.

There's undeniably a connection between the intensity of Arab anti-Semitism and the existence of the state of Israel. The Al-Azhar conference occurred within a year of the 1967 war in which Israel won a decisive victory over its Arab enemies—and, in the eyes of many Arabs, a humiliating one. The conference proceedings were full of references to the Israeli triumph over the Arabs as proof that Allah was expressing his displeasure with the lack of piety among Muslims.

Nonetheless, it was far from inevitable that the Arab-Israeli conflict should take on the powerful dimension of religious hatred that it has assumed since 1968. States and peoples clash for many reasons that have nothing to do with religion, in the Middle East as elsewhere. And a shared Islamic faith did not prevent wars from erupting between Iran and Iraq, Iraq and Kuwait, or Egypt and the Sudan. The transformation of the Arab-Israeli conflict into a religious war resulted from a series of historical steps that are easy to trace in retrospect.

The new theology codified at Al-Azhar was quickly espoused by the growing ranks of Islamic fundamentalists, sometimes called Islamists. The Islamist movement of the twentieth century had emerged in response not to any external Jewish enemy but primarily to the perception that Islam had lost its way in recent centuries, first during the period of Ottoman (Turkish) rule, then under the impact of Western imperialism. The goal of these fundamentalists was to purify Islam, to search for the authentic voice of the Islamic religion, and to urge their fellow Muslims to listen to that voice.

In this search for authenticity, they turned to the Qur'an and the history of *al-Rashidoon,* the "rightly guided ones," the first four caliphs after Muhammad, who ruled over a militant, ever-expanding, triumphant Muslim empire in the seventh century. They viewed the type of Muslim society presented therein as an eternal ideal.

Most of the efforts of the early fundamentalists had been devoted to cleaning the houses of their own Arab governments. Thus the Muslim Brotherhood, for example, established in Egypt in 1928, was locked in conflict with the pro-British Egyptian government through the 1940s and with Nasser's revolutionary socialist government in the 1950s. For the Muslim Brotherhood, a true Islamic state could not be run by outside imperialists, nor could it countenance an imported socialist program (despite its Islamic trappings).

In seeking to establish a pan-Arabic, pure Islamic state, the fundamentalists also had to deal with the presence of Jews in their society. In the early twentieth century there were still many Jews living in Arab lands, but their status had changed significantly since the preimperialist days of *dhimmi*-hood. Jews had generally embraced the Western newcomers to the Arab world. Under Western protection, they had cast off many of their traditional restrictions and humiliations and had benefited from education opportunities offered by both Christian missionary schools and by a network of schools established under the auspices of the Alliance Israèlite Universelle.

In addition, many Jews were increasingly attracted to Zionism. To the Islamic fundamentalists, the *dhimmi* status of Jews was non-negotiable, and Zionism, which sought to remove Palestine from the Islamic sphere, was intolerable. So despite their long history of living in Arab lands, Jews increasingly came to be viewed as outsiders and imperialists by the Muslim majority. Arab violence against Jews rose along with violence against imperialist agents, and for many of the same reasons.

It was in this environment that Islamists first started wrenching the story of Muhammad's conflict with the Jews from its historical context. It provided them with a model on which to base their opposition to the changing Jewish role in their society and the Jewish identification with the infidel West. As Zionist aspirations for a Jewish state of Israel grew and solidified, the universalization of the negative aspects of the Qur'anic view of the Jews became all the more attractive to the Islamists and their many sympathizers.

The founding of Israel in 1948, and worse, its victory in 1967, gave the Islamist view even greater appeal among many in Muslim society. So did the decline of the secularist governments that ruled in the Middle East following the retreat of the Western imperialist powers.

Even among the masses of people who lacked exposure to or understanding of the finer points of the Islamist position, ethnic or religious identification with Arab and Muslim hatred of Israel was very strong. Israel was an affront either to their religious sensibilities (as an occupier of land that had once been under Islamic control) or to their pan-Arabist ones (as the last outpost of the colonial West in the Middle East). The surviving secular governments in turn promulgated anti-Semitism to appease and draw support away from the Islamists, whose goal was still the overthrow of the secular governments to establish new Islamic regimes.

Thus the leaders of the Arab world have chosen for their own reasons to treat the conflict with Israel not mainly as a secular conflict over land or resources but as a religious one. It's a dangerous game for them to play. But it carries with it a number of benefits for the Arab leaders.

First, it enables Arab leaders to tap in to the power of ancient stereotypes and prejudices against Jews, using age-old canards imported from Europe to inflame the hatred of their peoples against the Jewish enemy.

Second, it provides a pretext for calling upon all their coreligionists around the world to support the Arab *jihad* (holy war) against Israel. Thus the energies and resources not just of the 300 million Arab Muslims but of over a billion Muslims around the world can be directed against the tiny Jewish state.

Third, and perhaps most important, by inflaming religious and ethnic hatred against an external enemy, it serves the classic purpose of scapegoating: it distracts the attention of Arabs from the failures of their own leaders. The corruption and incompetence of a Yasser Arafat, for example, are largely overlooked by a Palestinian nation that has been convinced that the source of its troubles is the tyranny of Israel and the Jews.

So today, thanks to the propaganda of a number of fundamentalist Islamic clerics, supported by Arab leaders in many countries, the Arab-Israeli conflict has been transformed from a nationalist struggle into a religious one. When Palestinian suicide bombers go out on their deadly missions, they wrap themselves not in the banner of the Palestinian Authority but in the green and white flag of Islam. When terrorists record videotapes to inspire their followers and frighten their opponents, they don't talk about demands for land or autonomy; they talk about religious martyrdom and about their wish to kill Jews.

All of this is relatively new. It's easy to forget that only a generation ago the mainstream Arab leaders tried hard to avoid the appearance of outright anti-Semitism. Even when they promoted a trade boycott in an attempt to strangle Israel economically, they stressed that this was a boycott against a country, not against Jews in general. In the last ten years, and especially since 2000, this has changed dramatically. Today in the Arab media and in the words of Arab leaders, there is no more differentiation between Jews and Israelis. The radical imams are not shy about calling on their followers to kill the Jews. Anti-Semitism is now completely, shockingly, out of the Arab closet.

SPREADING THE MESSAGE OF HATE

The Islamic fundamentalists were the first to adopt and promulgate the new theological attitude toward the Jews. For example, the Muslim Brotherhood popularized the notion that Jews were the first and most dangerous of the four horsemen of apocalypse, symbols of the corruption and decadence of modern society. In a children's supplement to the Brotherhood's *al-Da'wa'* publication in October 1980, an article entitled simply "The Jews" exhorted:

> Brother Muslim Lion Cub, Have you ever wondered why God cursed the Jews in his Book? . . . God grew weary of their lies. . . . They associated others with God, they were infidels. . . . Such are the Jews, my brother, Muslim lion cub, your enemies and the enemies of God. . . . Such is their particular natural disposition, = the corrupt doctrine that is there. . . . They have never ceased to conspire against their main enemy, the Muslims. In one of their books they say: "We Jews are the masters of the world, its corrupters, those who foment sedition, its hangmen!" . . . Muslim lion cub, annihilate their existence, those who seek to subjugate all humanity so as to force them to serve their satanic designs. . . .

Younger Islamist groups have also adopted this theological anti-Semitism. Sayyid Muhammad Husayn Fadlallah, the spiritual mentor of the Hezbollah terrorist group, says, "The struggle against the Jewish state, in which the Muslims are engaged, is a continuation of the old struggle of the Muslims against the Jews' conspiracy against Islam." The terrorist group Hamas, founded in 1987, includes in its covenant its belief that history will end with a Manichaean conflict between Muslims and Jews: "The resurrection of the dead will not come until Muslims will war with the Jews and kill them; until the Jews hide behind rocks and trees, which will cry, 'O Muslim! There is a Jew hiding behind me, come on and kill him!'"

Today this theology is echoed and reinforced every Friday in the sermons of radical *a'immah* and *'ulema* that are televised throughout the Middle East. One example comes from Palestinian television, which broadcast an imam preaching from a mosque in Mecca on October 24, 1997, proclaiming, "The Jews always set traps for the community of Muslims. . . . The Koran repeatedly warns against the traps and plots of the 'People of the Book.' They relentlessly scheme in all times and places, and this is what they do today and tomorrow against the Muslim camp."

A still more explicit example was broadcast on Saudi Arabia's TV1 television station on April 19, 2002, which featured Sheikh Abd Al-Rahman Al-Sudais preaching from the Al-Harram mosque in Mecca:

Read history and you will know that yesterday's Jews were bad predecessors and today's Jews are worse successors. They are killers of prophets and the scum of the earth. God hurled his curses and indignation on them and made them monkeys and pigs and worshipers of tyrants. These are the Jews, a continuous lineage of meanness, cunning, obstinacy, tyranny, evil, and corruption. They sow corruption on earth. . . . O Muslims, the Islamic nation today is at the peak of conflict with the enemies of yesterday, today, and tomorrow, with the grandsons of Bani-Quraydah, Al-Nadiri, and Qaynuqa [Jewish tribes in the early days of Islam]. May God's curses follow them until the Day of Judgment. . . . The conflict is exploding and magnifying, the exploitation and greed are increasing, and the indulgence in humbling Arabs and Muslims and their holy places has become very serious by the world rodents that have revoked pacts and agreements. Treachery, sabotage, and cunning dominate their minds and injustice and tyranny flow in their veins. . . . They cannot but remain arrogant, reckless, corrupt, and harmful. Thus, they deserve the curse of God, His angels, and all people.

Thanks to the power of today's electronic media, this kind of hatred in theological garb is able to find a huge worldwide audience.

TARGETING THE MASSES

Much of the hatred of today's Arab anti-Semitism is conveyed visually, through cartoons and other graphic images. The population of Egypt is only 15 percent literate; other nations in the region have comparable literacy rates. Thus many millions of Arabs get their impressions of the world through pictures,

New Israeli settlements in the Golan
Al-Ahali, April 26, 2000

whether broadcast on television or from newspapers posted on walls, fences, and notice boards around the cities.

Here are a few samples of the viciously anti-Semitic cartoons that routinely appear in the Arab press.

Al-Watan, April 6, 2003 (Qatar)

Al-Watan (March 17, 2002)

Akhbar al-Khalij, 6/6/2002 (Bahrain)

Al-Watan, May 13, 2003 (Qatar)

Al-Ahram (Syria) May 29, 2002

Notice the unembarrassed use of traditional Jewish stereotypes, derived from old Western models, as well as the evocation of classic anti-Semitic themes: the Jew as puppet master, controlling others for his own pernicious purposes; the Jew as capitalist pig, wielding bags of ill-gotten wealth in his quest for world dominance; the Jew as bloodsucker, draining life from the innocent.

Some of today's unprecedented wealth of Arab anti-Semitic literature is deliberately targeted at children. It couches bigotry in the form of lessons in religion or history, arguing that the Jews, who in the past were the greatest enemies of Prophet Muhammad, have resumed their role and are today the greatest enemies of modern Islam. Thus, just as the defeat and slaughter of the Jews of the Arab world by the Prophet Muhammad helped promote the rise of Islam and its conquests in the seventh and eighth centuries, the annihilation of Israel will bring into being a new golden era of contemporary Islam.

Here are some excerpts from Egyptian booklets for children with anti-Semitic themes and illustrations. First, from *The Wars of the Prophet: The War Against Bani Qainuqa* by Mas'ud Sabri, published in Cairo by Yanabi' (the Bani Qainuqa was a Jewish tribe living in the town of Medina in the Arabian peninsula; the tribe was expelled by Muhammad, whose wars against the Jews of Medina ended in its defeat and destruction):

When the Prophet, peace be upon him, arrived in al-Medina, he signed a pact with the Jews in order that all should live in peace and security. But the Jews are forever and everywhere Jews, they do not wish any good to the Muslims, and yearn to exile them from their land.

The Jews did not resign themselves, they persistently attempted to spread hate among the Muslims, they looked down on the verses of Allah [the Quran] and the Muslims.

The only way to eliminate the Jews is through Holy War [jihad] for the sake of Allah, because they are the most villainous among Allah's creatures. They will by no means leave the al-Aqsa mosque, unless Holy War [is waged against them] for the sake of Allah.

From *The Wars of the Prophet: The War Against Bani Nadhir* by Mas'ud Sabri, published in Cairo by Yanabi' (Bani Nadhir was another Jewish tribe in the town of Medina that was expelled from it by Muhammad):

The Jews of al-Madinah rejoiced about the outcome for the Muslims of the battle of Uhud [where the Muslims suffered extensive losses]. They showed hostility and hate against the Prophet, peace be upon him, and his allies. In addition, they made contact with the infidels in Mecca and planned together with them to murder the Prophet, peace be upon him, although the Prophet, peace be upon him, had signed agreements with them. . . . But the Jews are forever treacherous. For them there are no commitments and no agreements. The wickedness of the Jews—may Allah's curse be upon them—grew in force, when they made up their mind to murder the Prophet. But Allah rescued his Prophet, peace be upon him.

Finally, from *Islam and the Palestinian Problem* by Dr. Abdallah Nasih Alwan, published in 2001 in Cairo by Dar al-Salam:

No other nation in ancient and modern times has carried the banner of fraud, evil and treachery as has the Jewish nation. No other human race throughout history or from anywhere in the world has acted in such a cruel and corrupt manner and provoked such conflicts between nations as has the Jewish race.

[In] their [the Jews'] machinations in present times, at the beginning of the 14th century after *hijrah* [the Prophet's journey from Mecca to Madinah], the Jews—may Allah's curse rest upon them—have been using devious ways of conspiracy and deceit in order to achieve their aspirations and carry out their plans of establishing their rule over the world, and take control of the world's core powers. They are targeting three main objectives . . .

The first objective: spreading dissent among the nations . . .

The second objective: corrupting the faiths of the nations . . .

The third objective: founding the State of Israel, with Palestine as its center, and stretching from the Euphrates to the Nile. (23–24, 36)

Children who learn their history and religion from writings like these will grow up with an implicit belief in the evil and treachery of Jews, a belief they are likely to retain long after they've forgotten the specific sources from which they imbibed it.

POISON IMPORTED FROM THE WEST

Still, traditional Islamic images, however radicalized and distorted, do not nearly account for the variety of negative representations of Jews and Judaism in the twentieth-century Middle East and today. To explain fully the demonization of contemporary Jews in the Arab world, we must look to the West.

Since the 1960s the Islamists have been routinely applying to the Jews a heightened level of satanic rhetoric that includes the ancient blood libel (charges of ritual murder and poisoning) as well as the belief that powerful Jews in government and business are engaged in a hidden conspiracy to rule the world. As we've seen, these beliefs about the Jews are not indigenous to the Middle East. They originated in Europe in the medieval and early modern periods and were imported to the Middle East by European traders, missionaries, and occasionally even government officials in the imperialist nineteenth century.

The first major application of the blood libel in the Middle East, for example, occurred in the Damascus Affair of 1840, in which Jews were blamed for the disappearance of a Capuchin friar and his Muslim servant. The accusation of ritual murder in the Damascus Affair, like the majority of some twenty charges of ritual murder in the Middle East before the twentieth century, was made by Christians. Indeed, since its genesis in the Middle Ages, the blood libel had been invested with christological significance and linked to an alleged Jewish desire to continue the supposed ancient attacks on Jesus Christ and his followers.

In the twentieth century, however, the blood libel and the charge of ritual murder were recast by Muslims as merely another part of the Jewish religion, directed against not only Christians but Muslims and any other non-Jew as well. Thus "The God of the Jews is not content with animal sacrifices," wrote the Egyptian ʿAbdallah al-Tall in his 1964 book titled *The Danger of World Jewry to Islam and Christianity.* "He must be appeased with human sacrifices. Hence the Jewish custom of slaughtering children and extracting their blood to mix it with their *matzot* on Passover."

Similar blood libel accusations continue to appear in Arab media today. Arabic mass circulation newspapers in Qatar, Egypt, Saudi Arabia, Kuwait, Bahrain, and Jordan have reprinted similar claims about Jews and Israelis. And Arab political leaders have personally lent credence to the blood libel. In August 1972 King Faisal of Saudi Arabia reported in the Egyptian magazine *al-Musawar* that while he was in Paris "the police discovered five murdered children. Their blood had been drained, and it turned out that some Jews had murdered them in order to take their blood and mix it with the bread that they eat on that day."

Similarly, in 1984 Syrian defense minister Mustafa Tlass published a book, *The Matzah of Zion,* in which he returned to the Damascus Affair of 1840, claiming that the Jews had indeed murdered the Capuchin friar. In 2001 an Egyptian producer, Munir Radhi, announced that he was adapting Tlass's book into a movie. "It will be," he said, "the Arab answer to *Schindler's List*."

(A new variation of the blood libel involves accusations that the Jews are distributing poisoned water and food products in Palestine, as publicized by Sura Arafat, the wife of the Palestinian leader, in a November 1999 speech attended by First Lady Hillary Clinton.)

THE *PROTOCOLS* IN THE MIDDLE EAST

Another toxic Western import is *The Protocols of the Elders of Zion*, the notorious forgery that supposedly documents a Jewish plot to rule the world through treachery, fraud, and secret violence. The *Protocols* were discredited by the late 1920s, when the original text on which they were modeled—a tract from the 1860s attacking Napoleon III, with no mention of Jews—was discovered. Since then their influence in the West has been limited to the extremist fringe. But in the late 1920s the popularity of the book in the Arab world was only beginning, and, according to historian Bernard Lewis, its authenticity was not seriously questioned in the Arab media until the late 1970s.

Many Arabic translations of the *Protocols* are available, a good number of them published and republished by government presses in Egypt. By all indications they are perennial best-sellers in the Middle East. Until recently the *Protocols* was even on sale in the book-shop of the magnificent Western-style InterContinental Hotel in Amman, Jordan. (The book was removed from the shelves there after an ADL contingent staying at the hotel complained.)

The seriousness with which the *Protocols* appear to be taken in the Middle East may be explained partly by the number of prominent Muslims who have endorsed them. Egypt's President Nasser endorsed the *Protocols* in 1958, as did President Sadat, President

Arif of Iraq, King Faisal of Saudi Arabia, Colonel Qaddafi of Libya, and others. The *Protocols* form part of the worldview of extremist groups, as attested to by the mention by name in article 32 of the covenant of the terrorist group Hamas in describing the aspirations of Israel:

The Zionist plan is limitless. After Palestine, the Zionists aspire to expand from the Nile to the Euphrates. When they will have digested the region they overtook, they will aspire to further expansion, and so on. Their plan is embodied in the "Protocols of the Elders of Zion" ["Brutukulat Hukama Sahyun" in the Arabic original], and their present conduct is the best proof of what we are saying.

Excerpts from the *Protocols* have even appeared in the Jordanian school curriculum. Today the *Protocols* continue to be cited by public figures and in the media in the Arab world. In December 1997 Mustafa Tlass, the Syrian defense minister, cited the *Protocols* as an explanation for the warm relations between Israel and Turkey. On June 23, 2001, the Egyptian government daily *Al-Ahram* wrote:

What exactly do the Jews want? Read what the Ninth Protocol of "The Protocols of the elders of Zion" says: "We have limitless ambitions, inexhaustible greed, merciless vengeance, and hatred beyond imagination. We are a secret army whose plans are impossible to understand by using honest methods. Cunning is our approach, mystery is our way. [The way] of the freemasons, in which we believe, cannot be understood by those among the gentiles who are stupid pigs. . . . The ultimate goal of the free-masons is to destroy the world and to build it anew according to the Zionist policy so that the Jews can control the world . . . and destroy the [world's] religions. . . ."

The *Protocols* have also been cited in support of yet another variation on the blood libel, the bizarre assertion of Jewish or Israeli involvement in the attacks of September 11. Barely two weeks after September 11, 2001, a columnist in the Egyptian newspaper *Al-Wafd* wrote that the "Zionists" must have known in advance that the September 11 terrorist attacks were impending but refused to share that information with the United States "in order to sow disputes and troubles" throughout the world. "Proof is found," he added, "in the Protocols of the Wise Men of Zion."

Most recently, the *Protocols* emerged as a major plot element in the forty-one-part television series *Horseman Without a Horse*, which appeared on Egypt state television and other stations across the Middle East during the Islamic holy month of Ramadan.

In one episode, three stereotypical Jews with long gray beards and large black skullcaps are shown sitting in a room filled with religious symbols, including two Jewish candelabra. The elderly men are shown whispering about "the book" and expressing concerns that the *Protocols* may reach Egypt from Russia, where there are many copies.

In the scene, the Jewish characters express concern that a woman named Margaret has a copy of the book and is about to reach Egypt. The scene reinforces the anti-Semitic notion of the *Protocols* as a document that was created by Jews in a plot to control the world. The scene includes the following exchange of dialogue, where the characters discuss how to conceal "the truth" that the *Protocols* was written by Jews:

Ovad: Our group in Russia confirmed that all of the copies of the book *The Protocols of Elders of Zion* have been taken from the market.

Yitzhak: There is another edition of the book, Binyamin, from 1905, last year. In two days all the copies disappeared from the market. It is a false version.

Ovad: What about them? What can we do about that one book?

Yitzhak: The British know what the contents of the book are, but if the Egyptians find out there will be more trouble than there was in Russia. It can publicly expose our plans.

Binyamin: Our group can spread rumors throughout Egypt that the book is a forgery.

Yitzhak: Our problem is that someone will read the book and fit the contents to what we are trying to accomplish, and will believe in [its truth] thoroughly.

Even where the *Protocols* are not mentioned by name, the theme they express—that Jews are engaged in secret machinations to take over the world or that Jews already control the world—pervades the Arab worldview. A few examples of typical quotations from the mainstream Arabic press will illustrate the point:

At the end of the last century, the Jewish organizations consolidated a hellish plan to take over the world by sparking revolutions or taking control of the keys to governments in various countries, first and foremost the U.S. and Russia. . . . The Jewish sense of superiority is typified by hypocrisy and zeal. The Jews are incapable of actualizing their influence and control for a simple reason, and that is that they are a demographic minority in every society in the world. For this reason, the Jews are trying by means of their trickery to weaken the national identity [of the non-Jews] and thus take over affairs and direct them to serve their interests. (From the Saudi-government-controlled newspaper *Al-Watan,* December 8, 2001)

The Jews have been behind all the wars and their goal was corruption and destruction. This is their means of getting rich quick after wars. (From the Cairo weekly *Al-Ahram*, November 14, 1998)

Everywhere, the Jews have been the subjects of hatred and disdain because they control most of the economic resources upon which the livelihoods of many people are dependent. . . . There is no alternative but to say that the success of the Jews is not coincidental but rather the result of long years of planning and a great investment of effort in order to obtain their wretched control over the world's media. . . . (From the Palestinian Authority newspaper *Al-Hayat Al-Jadeeda*, July 2, 1998)

[Jewish] history is full of devising conspiracies, even against the countries in which they live, whose citizenship they bear and whose benefits they enjoy. . . . Anyone interested in documents from World War I can learn about the role German Jews played in organizing conspiracies to undermine Germany, harm its economy and weaken its capabilities, which deteriorated to the extent that it led to its defeat. Whoever studies these documents can also understand why the hatred of Jews consequently increased so severely. (From Damascus Radio, September 2, 1998)

In the United States sentiments like these can be found almost exclusively in the publications and Web pages of the most extreme hate groups. It's frightening to see them published routinely by the leading mainstream media outlets in the Middle East—fertilizing the growth of the next generation of plane hijackers and suicide bombers.

NAZISM AND HOLOCAUST DENIAL IN THE ARAB WORLD

Until recently, Arabs generally acknowledged the reality of the Nazi Holocaust. Of course, Arab anti-Semites expressed little sympathy for the Jews concerning the Holocaust. In fact, their dominant attitude was one of resentment. Because they regarded the establishment of Israel in their midst as an attempt by the world community to recompense the Jews for the horrors they suffered, Arab anti-Semites often complained, "Why should we have to suffer for crimes that the Europeans committed?"

Now this attitude is changing, as many Arab anti-Semites are embracing Holocaust denial.

Of course, Holocaust denial is not an exclusively Arab problem. Like most other elements of global anti-Semitism, it originated in the West. A network of "historians," some with seemingly impressive credentials, generates a steady stream of documents supposedly demonstrating that the Holocaust never happened or, if it did, that it was far less serious than generally believed. Though the deniers often try to portray themselves as revisionists conducting legitimate inquiry into the historical record, scratching the surface of their theories demonstrates the anti-Semitic conspiracy theories they are really espousing.

To make their claims credible, Holocaust deniers are forced to reject enormous volumes of historical evidence from World War II, resorting to fantastic conspiracy theories about Jews. Records from the period, including thousands of pages of evidence used immediately after the war in the Nuremberg trials, are dismissed as forged by a secret committee; survivors are rejected as greedy charlatans; American GIs who saw the death apparatus in the camps are said to have been duped by the American military itself, which was corrupted by Jewish concerns and also complicit in the conspiracy. The implausibility of it all would make these notions comical, if the subject weren't such an important and tragic one.

As for a motive, deniers claim that the Jews wanted to defraud the West of billions of dollars in reparations and other payments; to purchase world support for the creation of the state of Israel;

and to demoralize "Aryans" and the West so that the Jews could more easily take over the world.

Although Holocaust denial originated in the West, it has found a comfortable home in much of the Arab world. Holocaust denial now regularly occurs throughout the Middle East, in speeches and pronouncements by public figures, in articles and columns by journalists, and in the resolutions of professional organizations. Even as many Western countries have enacted legislation identifying Holocaust denial as a form of hate crime, governments in the Middle East do not condemn, and sometimes even sponsor, such anti-Semitic propaganda.

For example, Iran has become a sanctuary for Western Holocaust deniers fleeing legal entanglements in their home countries, and its leader, Ayatollah Ali Khamenei, suggested in 2001 that Jewish deaths during the Holocaust had been exaggerated:

> There is evidence which shows that Zionists had close relations with German Nazis and exaggerated statistics on Jewish killings. There is even evidence on hand that a large number of non-Jewish hooligans and thugs of Eastern Europe were forced to migrate to Palestine as Jews. The purpose was to install in the heart of the Islamic world an anti-Islamic state under the guise of supporting the victims of racism and to create a rift between the East and the West of the Islamic world. (*Jerusalem Post*, April 25, 2001)

The many expressions of Holocaust denial that have appeared in *Teshreen*, Syria's main daily newspaper, which is owned and operated by the ruling Baath party, show that the Syrian government also condones the propaganda. The same holds true for the Palestinian Authority, whose newspaper, *Al-Hayat Al-Jadeeda*, and television station have frequently denied basic facts of the Holocaust in their reporting.

In other Middle Eastern countries, denying or minimizing the extent of the killing of Jews during World War II has been adopted by opposition parties and dissident factions that oppose attempts at normalizing relations—legal, diplomatic, economic—with Israel or the United States. For these factions, Holocaust denial is a tool to discredit their government rivals, who have allegedly been taken in by Israeli Holocaust propaganda, and to increase popular hatred of Israel. This is true in Jordan, for example, where antinormalization organizations sought to hold Holocaust-denial conferences in 2001 but were opposed by the Jordanian government. (Nonetheless, the strongly anti-Zionist Jordanian Writers' Association was able to hold its conference.) The Lebanese government also opposed the attempts of several foreign organizations to hold a Holocaust-denial conference in Beirut in 2001.

Arab attempts to deny or minimize the Holocaust date back to the 1980s. For example, in the late 1980s a former Moroccan army officer, Ahmed Rami, fled to Sweden after being sentenced to death in Morocco for his role in a 1972 coup attempt against King Hassan II. There he founded Radio Islam, an anti-Semitic program that featured attacks on accepted Holocaust history. In October 1990 Swedish courts ruled that Rami and Radio Islam were guilty of incitement against Jews; Rami received six months in prison, and Radio Islam was shut down.

Two years later Rami was a featured speaker at the annual conference of the Institute for Historical Review, the leading Holocaust-denying organization in the United States. He eventually reestablished Radio Islam as an Internet site but was again prosecuted in Swedish courts on charges of inciting racial hatred and was convicted and fined in October 2000. Radio Islam continues to maintain its Web presence from servers in the United States; though it no longer makes audio broadcasts, its site features an extensive collection of Holocaust-denial and other anti-Semitic documents.

Since the 1990s Holocaust denial has become increasingly popular in Arab media throughout the Middle East. This is true even in Egypt and Jordan, the two Arab countries that have taken steps to normalize relations with Israel. Among the newspapers that have consistently featured Holocaust denial are the Jordanian daily, *Al Arab Al-Yom,* the Syrian daily, *Teshreen,* the English language Iranian *Tehran Times,* and the Palestinian Authority's *Al-Hayat Al-Jadeeda.*

Several noted religious leaders in the region have also rejected the facts of the Holocaust, including Sheikh Mohammad Mehdi Shamseddin of Lebanon, Sheik Ikrima Sabri of Jerusalem, and Iranian religious leader Ayatollah Ali Khamenei. On July 4, 1998, for example, the establishment Jordanian newspaper *Al-Arab Al-Yom* told its readers that "most research prepared by objective researchers" has "proven in a manner beyond the shadow of a doubt" that the Holocaust is "a great lie and a myth that the Zionist mind spread in order to lead the world astray." Earlier that year (April 27, 1998) the same newspaper published an article claiming that "there is no proof" that the Holocaust occurred, except for "the conflicting testimonies of a few Jewish 'survivors.'"

Similarly, on July 14, 1998, the Egyptian newspaper *Al-Akhbar* stated that regarding the crematoria remaining at Buchenwald and Auschwitz, "even if these crematoria operated day and night, it would take dozens of years to burn six million people and not merely three years." A Lebanese politician, Dr. Issam Naaman, wrote in a London-Arab newspaper on April 22, 1998, that "Israel prospers and exists by right of the Holocaust lie and the Israeli government's policy of intentional exaggeration."

Several Western Holocaust deniers have turned to the Arab and Islamic world for help when facing prosecution in various countries for illegal activities. Wolfgang Fröhlich, an Austrian engineer who testified on behalf of Swiss denier Jurgen Graf in 1998 about the supposed impossibility of Zyklon-B gas being used for executing humans, sought refuge in Iran in May 2000, claiming that his arrest by Austrian police was imminent. He reportedly still resides in that country.

Graf himself, who was convicted of inciting racial hatred by promulgating Holocaust denial in Switzerland in 1998, also resides in Iran, to which he fled rather than face a fifteen-month jail term. According to the Institute for Historical Review, Graf is presently living in Tehran "as a guest of Iranian scholars." Since his arrival in Iran, Graf has authored an online book titled *Holocaust Revisionism and Its Political Consequences.* The relationship between either Fröhlich or Graf and the Iranian government is not clear.

The best-known flare-up of Holocaust denial in the Middle East occurred in response to the trial of Roger Garaudy in France in 1998. Garaudy was charged with violating a 1990 French law that makes it illegal to deny historical events that have been designated as "crimes against humanity" and with inciting racial hatred. These charges stemmed from his 1995 book, *The Founding Myths of Modern Israel* (*Les mythes fondateurs de la politique israélienne*), in which he stated that there was no Nazi pogrom of genocide during World War II and that Jews essentially fabricated the Holocaust for their financial and political gain. Garaudy was convicted on these charges in 1998.

Before, during, and after the trial, he was hailed as a hero throughout the countries of the Middle East; the trial was covered by media from Saudi Arabia, Qatar, Egypt, Iran, Syria, Lebanon, Jordan, and the Palestinian Authority. Formerly Roman Catholic and Communist, Garaudy had converted to Islam in 1982 and married a Jerusalem-born Palestinian woman, but this alone did not explain the outpouring of support he received; the "revisionist" message of his book, whose Arabic translation was a bestseller in many of the region's countries, clearly resonated across the region. The former president of Iran, Ali Akbar Hashemi

Rafsanjani, announced in a sermon on Radio Tehran that his personal scholarship on the subject had convinced him that "Hitler had only killed twenty thousand Jews and not six million," and added that "Garaudy's crime derives from the doubt he cast on Zionist propaganda."

The main establishment newspaper in Egypt, *Al-Ahram,* defended Garaudy in a March 14, 1998, article that argued that there is "no trace of the gas chambers" that are supposed to have existed in Germany and that six million Jews could not have been killed in the Holocaust because "the Jews of Germany numbered less than two million" at the time. Numerous professional and social organizations throughout the region issued statements supporting Garaudy as well, including the Palestinian Journalists' Syndicate, the Palestinian Writers Association, the Jordanian Arab Organization for Human Rights, the Qatar Women's Youth Organization, the Federation of Egyptian Writers, and the Union of Arab Artists.

Support for Garaudy did not end merely with words. Seven members of the Beirut Bar Association volunteered to defend the writer in France, and Egypt's Arab Lawyers' Union also dispatched a five-man legal team to Paris in Garaudy's support. The United Arab Emirates daily, *Al-Haleej,* was inundated with contributions and messages of support for Garaudy after it published an appeal on his behalf. The most surprising contribution came from the wife of United Arab Emirates leader Sheikh Zayed ibn Sultan al-Nahayan, who gave the equivalent of fifty thousand dollars in cash to cover the maximum fine that Garaudy would be required to pay if found guilty.

Another aspect of Arab attitudes toward the Holocaust is the perception that the West created Israel out of guilt over its attempted genocide of the Jews. This leads many Arabs to complain that they are paying for the sins of the West by being forced to accommodate the existence of Israel. This opinion was frequently voiced by Palestinian opinion makers—until the break-down of the 2000 Palestinian-Israeli peace process, when many came to view the recognition of *any* historical Jewish suffering as a political liability, and the Palestinian Authority—controlled media outlets increased their dissemination of Holocaust denial.

This change in emphasis suggests that the issue is not primarily the depth of anti-Semitism but the reduction of inhibitions about expressing such extreme forms of hatred. The Palestinians no longer merely have to talk about their suffering as a consequence of Jewish suffering; now they can get away with accusing the Jews of being the perpetrators of a fraud.

Another troubling approach to the Holocaust also exists in the Middle East. Hatred of Israel has led some Arabs to embrace Nazism itself and to applaud its attempted genocide of the Jews. "[Give] thanks to Hitler," wrote columnist Ahmad Ragab recently in the Egyptian newspaper *Al-Akhbar*. "He took revenge on the Israelis in advance, on behalf of the Palestinians. Our one complaint against him was that his revenge was not complete enough."

At the same time, in a perverse twist, Arab opinion makers have consistently used World War II—era associations to describe Israel and its actions. Israeli leaders have been compared with Hitler and its army with the SS; Palestinian refugee camps have been dubbed concentration camps. Associating Israel with Nazi Germany in general remains a standard rhetorical device. But, especially of late, propaganda demonizing Israel as Nazi competes in Arab media and politics with propaganda that denies the existence of the Nazi Holocaust.

An example of this contradiction—condemning Israel with Nazi labels while denying the worst of the Nazi crimes—can be found in the Syrian daily, *Teshreen*, on January 31, 2000. In the space of a single column ("The Plague of the Third Millennium"), editorialist Muhammad Kheir Al-Wadi called on the international community to "adamantly oppose the new Nazi Plague that breeds in Israel" while claiming that Zionists "invented" the notion of a "Nazi Holocaust in which the Jews suffered." Only blinding hatred could induce a self-respecting writer to utter such self-contradictory nonsense.

JIHAD ONLINE

Like American extremists, Islamist terrorists and their supporters use the Internet to communicate covertly, to preach to the public, and to solicit funds. They also use it to plan and coordinate their attacks, such as those of September 11. In fact, there is evidence that they are planning to turn the Internet itself into a terrorist weapon, using it to wreak havoc on America's critical infrastructure.

For example, operatives of Al-Qaeda, the leading multinational Islamist terrorist network, relied heavily on the Internet for help in planning and coordinating the September 11 attacks. Thousands of encrypted messages that had been posted in a password-protected area of a Web site were found by federal officials on the computer of arrested Al-Qaeda operative Abu Zubaydah, who reportedly masterminded the September 11 attacks. The first messages found on Zubaydah's computer were dated May 2001, and the last were sent on September 9, 2001, two days before the attacks. The frequency of the messages was highest in August 2001, the month immediately preceding the attacks.

Both online and off, Al-Qaeda members operate clandestinely and are difficult to deter. To preserve their anonymity, they use the Internet in public places and send messages via e-mail accounts that were likely registered using fabricated information. In addition, these extremists use the Internet to research their targets and weapons of choice. Most significantly, they use sophisticated computer programs to stop anyone but their compatriots from reading or finding the messages they transmit online.

Some of the September 11 hijackers accessed the Internet in public libraries. It is often simple to use the Internet in such facilities without being traced or identified; at many public libraries nearly anyone can walk up to a terminal and log on to the Internet without presenting identification. Another place Al-Qaeda members have accessed the Internet is in *hawalas,* storefront money exchanges that have also been used to funnel money to bin Laden and Al-Qaeda. The United States government raided and shut down dozens of these *hawalas* in November 2001.

Al-Qaeda operatives have used the Internet to search for and find logistical information to use in planning attacks. Some of the September 11 hijackers used the Internet to research the chemical dispersal capabilities of crop dusters, and their ringleader, Mohammed Atta, made his plane reservations online. Information about manufacturing a nuclear bomb and files describing American utilities, downloaded from the Internet, were found in Al-Qaeda safe houses by American officials in Afghanistan.

A key tool of Al-Qaeda operatives has been encryption programs, which scramble messages so they cannot be read. Only the intended recipients of an encrypted message will be able to read it, using a special electronic decoding key received from the sender of the message. Easy-to-use encryption programs are readily available, free of charge, on the Internet.

Encryption became so important to Al-Qaeda after the United States began to intercept bin Laden's satellite telephone calls that the topic was added to the curriculum of the group's terrorist training schools in Afghanistan and Sudan. Virtually uncrackable encryption allows Al-Qaeda members "to communicate about their criminal intentions without fear of outside intrusion," former FBI director Louis Freeh told a Senate panel. "They're thwarting the efforts of law enforcement to detect, prevent, and investigate illegal activities," he warned.

Al-Qaeda also reportedly uses the Internet to disseminate so-called steganographic messages, secret statements or pictures embedded in apparently harmless information that is posted publicly online. Al-Qaeda operatives have been accused of hiding attack plans in apparently commonplace image files available in sports chat rooms and on pornographic Web sites. (Ironic, isn't it, that Islamic extremists, who are supposedly devoted to a puritanical lifestyle, would use pornography to transmit their attack plans?) It's generally possible to detect a steganographic message in an online photo by analyzing the file in which the image is stored. Dozens of government analysts have been searching for, finding, and working to decode these messages since September 11.

Some intelligence agents assert that Al-Qaeda operatives post logistical information about upcoming attacks on the Web in plain sight, without hiding the information in photos. These analysts assert that Al-Qaeda operatives, but not others, are privy to the meaning of certain signals, such as phrases and symbols, that appear on sites sympathetic to the group. Officials have noticed that traffic at these sites surges before attempted attacks. This type of rudimentary steganography resembles the kind used by British radio announcers during the Second World War, who stealthily alerted the French Resistance of upcoming attacks by using seemingly nonsensical, prearranged phrases in their broadcasts.

Finally, Al-Qaeda operatives have investigated using the Internet itself as a weapon. Some Al-Qaeda computers seized by American forces in Afghanistan contained information about ways to gain control of the computers that run electrical, water, transportation, and communications systems. Many of these computers are accessible over the Internet, and security is lax. With access to such computers, Al-Qaeda operatives could potentially shut off the electricity of a city, shut down the phone lines in a given neighborhood, cause a dam to release the water it is holding, or cause two trains to crash into each other.

Many government experts assume that Al-Qaeda operatives will actually launch a cyberattack of this sort in conjunction with a more conventional attack. For example, terrorists could blow up a building and then disable the phone system in the surrounding area in order to prevent law enforcement, medical, and emergency officials from responding to the attack.

Virulent anti-Semitism is an integral part of Al Qaeda's message. In a statement published June 2, 2002, in the pan-Arab newspaper *Al-Hayat,* Al Qaeda spokesman Sulaiman Abu Gaith warned that new attacks against "Americans and Jews" were coming. The reason for the linkage, he explained, was that the United States is "in partnership with Jews, the head of corruption and decay . . . the reason behind all the injustice and oppression that befell on Muslims." The war on terror is a matter of grave importance for every American; as remarks like this make clear, it carries special importance for Jewish-Americans.

ISLAMIST PROPAGANDA AND FUND-RAISING ONLINE

The online propaganda strategy of Al-Qaeda takes advantage of the anonymity and flexibility of the Internet, relying on semiofficial sites instead of official sites to spread its message. These sites have been the first to publish statements apparently composed by Al-Qaeda officials and allies, including bin Laden and Taliban leader Mullah Omar.

The people who maintain such sites could be Al-Qaeda members, supporters of Al-Qaeda who are in direct contact with its members, or simply like-minded Islamists. Like the Al-Qaeda members who communicate with each other through e-mail, they are difficult to identify. By using fictitious information, they can easily register a Web site or create an e-mail account.

All of the semiofficial Al-Qaeda sites except for that of Azzam Publications, a British Islamist bookseller, contain only

Arabic-language materials, but these sites are hosted throughout the world. Many are registered or hosted in Europe (Italy, Sweden, England, Poland), Asia (Malaysia), or the United States. In addition to statements from bin Laden and Al-Qaeda, these sites feature, among other items, articles that condemn America, biographies of Islamists killed in battle, and biased accounts of current news.

For example, an article on the Al-Maqdese site, titled "An Urgent Call to the Diabolical Americans," asks Muslims to "curse" President Bush and Americans with "immediate perdition." "I personally prayed to God for it," the author boasts, "because the Muslims and I know that you are from among the most impure, unbeliever and abject peoples in the world. You are standing behind every heresy and crime, and behind every shame, impudence and trouble. You are the commander of the religious and moral weakening."

At the Al-Emarh Web site, which promotes the Taliban, Mulla Omar states, "The Muslims know that America only wants to fight Islam and to liquidate everyone who acts according to the Islamic Shariah, because America knows that the biggest danger to it and for the Jews is Islam and its believers." America, he claims, "wants to rule the world and to eat the wealth and properties of the weak people."

The Azzam Publications site features more than four dozen celebratory biographies of "Foreign Mujahideen Killed in Jihad." Most of these accounts were written by personal associates of the deceased militants, and many of the stories are accompanied by photos, audio clips, or video footage. These laudatory reports, most of which concern extremists killed in Afghanistan, Chechnya, or Bosnia, clearly intend to inspire prospective recruits to take up arms.

Al-Qaeda and its supporters provide interested parties around the globe with their view of the conflict in Afghanistan through sites such as jehad.net. That site, which describes itself as "a news network" covering "the latest events" and all of the issues related to "the Crusade war" in Afghanistan, complains about alleged American atrocities against Afghan civilians while cheering allegedly successful attacks by Al-Qaeda and the Taliban on American forces. Sites supporting Al-Qaeda have consistently claimed death tolls for American troops hundreds of times larger than those reported in the Western press, giving readers the impression that Al-Qaeda and its allies are holding their own, if not prevailing, in their war against the United States.

Al-Qaeda has also received funds collected over the Internet by seemingly legitimate charities. The government so far has frozen the assets of three charities that use the Internet to raise money—the Benevolence International Foundation, the Global Relief Foundation, and the Al-Haramain Foundation—because of evidence that those charities have funneled money to Al-Qaeda.

For example, the Benevolence International Foundation (BIF), based in Illinois, describes itself on its Web site as "a humanitarian organization dedicated to helping those afflicted by wars." The site provides a bank account number where donations can be wired, information about donating by credit card, and a form that can be used to initiate monthly donations through automatic withdrawal from the donor's bank account. In addition, the BIF site offers information about corporate matching-gift programs and instructions on how to donate stocks.

BIF, which has raised millions of dollars a year, transferred money on behalf of bin Laden in the 1990s, and as recently as April 2000 it sent more than $600,000 to Chechnyan extremists trained by Al-Qaeda. The charity arranged a trip to Bosnia for Mamdouh Salim, who traveled to Sudan in 1994 to obtain radioactive material for Al-Qaeda. Mohamed Bayazid, another man suspected of trying to acquire radioactive material for Al-Qaeda, listed the BIF address as his home address on his driver's license. BIF has also been linked to Islamists involved with the 1993 World Trade Center bombing.

Al-Qaeda isn't the only terrorist organization that relies on the power of the Internet. Hamas and Palestinian Islamic Jihad (PIJ), the two leading Islamist terrorist groups operating in the Palestinian territories, have long used the Internet to plan and coordinate their operations and to provide practical advice to terrorist operatives. "We use whatever tools we can—e-mail, the Internet—to facilitate jihad against the occupiers and their supporters," claims Hamas founder and leader Sheik Yassin. "We have the best minds working with us." Israeli intelligence believes that Hamas uses the Internet to transmit maps, photographs, directions, and codes related to specific attacks, plus nuts-and-bolts guidance on how best to detonate a bomb.

HEZBOLLAH: "PSYCHOLOGICAL WARFARE AGAINST THE ZIONIST ENEMY"

Yet another terror group that makes heavy use of the Internet is Hezbollah, also known as the Party of God, the leading Islamist terrorist group in Lebanon. Founded in 1982 by Islamist clerics under the influence of Iran, Hezbollah, which is supported financially by Iran and Syria, enjoys an annual budget of approximately $50 to $100 million. The group has engaged in both political and military activities, winning seats in the Lebanese parliament while launching thousands of attacks on Israeli and American civilian and military targets. In the 1980s Hezbollah kidnapped American citizens, and in 1983 the group brutally murdered more than two hundred American servicemen by bombing their barracks in Beirut. Hezbollah is also alleged to have killed more than a hundred people in bombings of the Israeli Embassy (1992) and a Jewish community center in Argentina (1994).

Hezbollah has established a large network of linked Web sites, some with versions in English and others available only in Arabic. The official English-language Hezbollah Central Press Office site features public statements from the group, transcripts of speeches given by its leader, Hassan Nasrallah, songs celebrating *jihad,* and a propaganda film. The introductory article on the site is full of contradictions. It claims that Hezbollah is "far away" from "any fanaticism," wants to "establish peace and justice to all humanity whatever their race or religion," and "stretches its arm of friendship to all on the basis of mutual self-respect." At the same time, it notes that Hezbollah uses "its own special types of resistance against the Zionist enemy," namely, "suicide attacks," which "dealt great losses to the enemy on all thinkable levels such as militarily and mentally" and raised morale "across the whole Islamic nation."

The Arabic version of the site contains more content than the English version, and unlike the English version, it repeatedly refers to Israelis as Nazis.

Hezbollah's Islamic Resistance Support Association site complements the Central Press Office site. Throughout this site, the term *Israel* appears in quotation marks, reflecting the belief of Hezbollah that no legitimate state of Israel actually exists. The site documents the activities of the group with videos of its rallies and military operations, daily reports describing its attacks on Israeli targets, and an encyclopedia of its "martyrs"—suicide bombers and others killed in the fight against Israel—complete with photos of them and information about their families.

In order to justify its violent activities, Hezbollah posts "legal" documents on the site that validate "armed resistance." One Arabic article on the site calls the verses of the Torah, Judaism's holy book, "a fundament in the building of the hate of the others." The site also borrows propaganda from Western anti-Semites, reprinting the writing of French Holocaust denier Roger Garaudy and deceased American neo-Nazi William Pierce.

Another Hezbollah site offers content from Al-Manar, the satellite and cable television station owned and operated by the group. Established in 1991, Al-Manar describes itself as "the first Arab establishment to stage an effective psychological warfare against the Zionist enemy." Thousands of Palestinians watch Al-Manar, which has an annual budget of approximately $10 million. The Al-Manar Web site presents video broadcasts of, and text based on, the station's English-language news broadcasts. One of the text documents found there is the source of the lie that four thousand Jews chose not to come to work at the World Trade Center on September 11 because they knew of the attack in advance.

The site for the Hezbollah magazine *Baqiyyat-u-llah,* which is completely in Arabic, promotes *The Protocols of the Elders of Zion,* as does the site for the Al-Quds Cultural and Social Association, which is also tied to Hezbollah. According to the Al-Quds site, "the Jews see" the *Protocols* "as a basis for their behavior and as a place to keep their plans," and the *Protocols* expose "the reality of the Jewish soul, which does not carry anything except corruption." That site recommends that its readers read the *Protocols* "to know the Jews' racial and destructive way of thinking."

On its Al-Manar Web site, Hezbollah urges visitors to donate funds "for the sustenance of the Intifadah." The site specifies three particular bank accounts in Lebanon to which donors can send money. Hezbollah also operates three Web sites for charities that help support its military operations: Al-Shahid and Al-Emdad, which fund the families of Hezbollah soldiers killed in battle, and Al-Jarha, which funds wounded Hezbollah soldiers. Undoubtedly, Hezbollah recruits are probably more willing to engage in terrorist attacks knowing that they will be cared for if they are injured and that their families will be taken care of if they are killed.

ARAB ANTI-SEMITISM: LAND MINE ON THE
ROAD TO PEACE

Obviously, the epidemic of anti-Semitism in the Arab world has become a significant obstacle to any lasting Middle Eastern peace, especially after decades of being deliberately spread by regional leaders and explicitly linked to the revered teachings of Islam.

This is not to say that Arab anti-Semitism can't be combated. To the extent that it expresses the political frustration long suffered by the peoples of the Arab world, it can be mitigated through resolving the underlying tensions. Now that the United States is taking the lead in pushing the Arab world toward democracy and freedom, perhaps the Muslim peoples can look forward to one day living in societies that provide their citizens with greater hope and opportunity, which may reduce their psychological need for a scapegoat.

Anti-Semitism based on religion, however, may be different. The constant trumpeting of the anti-Semitic Islamist paradigm since the late 1960s, and the apparent absence of an opposing, moderate voice of Islam emanating with any real force from the Middle East, suggests that this new, theologically based demonization of Jews is being accepted and internalized by masses of Muslims throughout the region. Because it weaves together a theological hatred of Jews with the worst of Western anti-Jewish conspiracy theories, today's anti-Semitism in the Middle East may not be easily overcome. Religions are self-propagating, and theologies possess remarkable staying power across generations.

We mustn't underestimate the significance of the anti-Semitic hate industry in the Arab world and the damage it causes. The virus of anti-Semitism doesn't remain confined to mosques and university campuses. It reaches far beyond the circles of dubious intellectuals and extremist clerics. It enjoys widespread circulation and popularity throughout the Arab and Muslim world as well as the Muslim communities in Europe and the Western world. In Britain, France, the United States, and many other countries, Islamic organizations raise funds for terrorists and promote hatred through preaching, publications, and Web sites. According to Professor Abdul Hadi Palazzi of the University of Vellectri in Rome, "Over 80 percent of European mosques are controlled by extremists who belong to radical pseudo-Islamic movements that have absorbed anti-Semitic motifs."

As we've seen, this hate industry constitutes an effective means of mobilizing popular support in the hands of dictators striving to channel the bitterness and frustration felt by their impoverished and suffering populations. It also serves as a valuable tool in the Israeli-Arab and Israeli-Palestinian struggle, and it reflects the basic unwillingness to accept the existence of the state of Israel and the challenge posed by the Zionist Jewish state in the very midst of the Arab and Muslim world.

Today the Arab hate industry uses anti-Semitism to encourage Palestinians and other Muslims to engage in murderous terrorist attacks against Israelis and Jews. It points to the "Jewish danger" as the main threat to Islam, to which holy war (*jihad*) is the only response. Potential terrorists are nurtured by anti-Semitic propaganda that portrays the Jews as descendants of apes and pigs, worshipers of the calf and of Satan, and cursed, unclean, and impious people whose greatest ambition is to desecrate mosques and murder innocent Muslims. Some suicide bombers have been found to have kept copies of *The Protocols of the Elders of Zion* and were obviously convinced they were conducting a struggle against a Jewish world-embracing conspiracy that poses a direct threat to the Muslim nations.

In the Middle East, even more than in the rest of the world, hatred leads directly to death.

TAKING A STAND

It shocks many Americans to realize that some countries that are supposedly staunch allies of the United States in the Middle East, such as Egypt and Jordan, are among those that promulgate anti-Semitic hatred. Perhaps you wonder why our government doesn't take a stronger stand against this bigotry.

In my decades of monitoring this problem, I've learned that the problem of Arab anti-Semitism rises and falls on the agenda of the United States government depending on various factors. The amount of pressure the United States is willing to exert on its Arab allies varies from time to time as policies and circumstances shift.

Our government does issue protests to our allies when egregiously anti-Semitic actions take place. For example, when Egypt publicized the use of the *Protocols* in its *Horseman* TV series, Secretary of State Powell raised the issue, as did our ambassador in Cairo. Congress has passed resolutions and threatened to cut off aid to countries that continue to promote bigotry. Still, these efforts aren't nearly as forceful and consistent as I would like.

For many years, even Israelis and American Jews paid little attention to Arab anti-Semitism. Our focus was on trying to start an Arab-Israeli peace process. Now we are learning that hateful attitudes make peace almost impossible—that gestures of reconciliation are quashed when they occur in a surrounding atmosphere of unrelenting hatred.

I once raised this issue with Israeli leader Ehud Barak. "Mr. Prime Minister," I said, "I don't understand how you can negotiate peace with an Arab enemy who is allowing the flames of hatred to be fed by the press every day. Can't you demand a crackdown on Arab anti-Semitism as a precondition of peace talks?"

Barak's reply was instructive. "My predecessor, Benjamin Netanyahu, felt as you do. He didn't want to make peace with the

Arabs, so he made lowering hatred a precondition. But I think differently. I believe you make peace first, and change attitudes later."

Barak was wrong, and he has publicly acknowledged a change of heart about this. He now views the problem as I do: that no true peace between Israel and its neighbors will be possible until the Arab press ceases its drumbeat of anti-Semitic propaganda.

Egypt claims to be the leader of Arab world, and maybe it is. It is also a close ally of the United States and a recipient of enormous amounts of American foreign aid (second only to Israel, in fact). Finally, it is supposedly at peace with Israel, having signed peace accords that include a promise to fight the spread of anti-Semitism. So when the Egyptian government started tolerating the publication of anti-Semitic diatribes shortly after President Hosni Mubarak came to power in 1981, I went to see him.

At first Mubarak tried to dismiss my complaints. He said, "Egypt is a democracy. I can't tell my people what to write, print, or read."

This was disingenuous. "With all due respect, Mr. President, Egypt is at best a controlled democracy. Editors are hired and fired at the behest of the government. If you told them that anti-Semitism is against the spirit of Islam and immoral, and behaved as if you really meant it, they would stop printing such stuff."

Mubarak responded by changing his argument. "Well, these writings and pictures you complain about are not really anti-Semitic; they are merely anti-Israel."

"Mr. President," I replied, "I understand the difference. And your newspapers contain plenty that is both anti-Israel *and* anti-Semitic."

Mubarak was about out of patience with me. He said, "I'm opposed to anti-Semitism. You tell them I said so." Our conversation was over. I subsequently held a press conference at which I quoted Mubarak's words—with no visible effect on the Egyptian press.

We've raised the same issue several times at subsequent meetings, and we'll continue to do so. We've pushed President Mubarak on other fronts as well. For example, the only time he has visited Israel was for the funeral of Prime Minister Rabin. If Egypt and Israel are truly at peace, surely he could make a second visit. When I told him so, President Mubarak replied, "You don't like me very much, do you, Mr. Foxman? Don't you know that visiting Israel could get me killed?"

Here, of course, is the heart of the problem faced by Mubarak and other so-called moderate Arab leaders. Having tolerated and encouraged the Islamists as a way of deflecting anger away from themselves and toward Israel and the West, they now find themselves riding the tiger of increasingly virulent fundamentalist rage. Their position is an increasingly dangerous one. The only real solution, of course, would take enormous courage: to stand up on behalf of true democracy, progress, tolerance, and freedom.

Someday an Arab leader will take such a stand. The one who does will ultimately be recognized as a figure of historic greatness, honored in the same way as Martin Luther King Jr. and Nelson Mandela. But who will do it? And when?

While the world waits, the hatred continues to spew forth from the mosques, newspapers, and TV studios of the Arab world, with incalculable consequences. A theologically based anti-Semitism gripped Christian Europe for hundreds of years; it took centuries of reformation and revolution and an all-encompassing reordering of society for its domination to be broken. Even today its aftereffects linger. I fear that the Middle East will require a process of change that is every bit as wrenching and difficult as the one Europe experienced.

What this tells me is that it is vitally important for nations and leaders around the world to take seriously these manifestations of hatred and denounce them. The United States and the European Union must not allow Arafat, Mubarak, and other leaders in the

Arab world to think that they can tolerate or encourage anti-Semitism in their societies with impunity.

SEARCHING FOR A PEACE PARTNER

As I write these words in mid-2003, a new glimmer of hope for peace in the Middle East has emerged. Within the last few days Israeli prime minister Ariel Sharon has convinced his cabinet to support (with reservations) the so-called road map for peace developed and backed by four international partners: the United States, Russia, the European Union, and the secretariat of the United Nations ("the Quartet"). Plans are being developed for talks with the newly installed Palestinian prime minister, Mahmoud Abbas, aimed at turning the road map into reality.

I'm always grateful and optimistic whenever the prospects for peace appear to be improving. But like the Israeli leaders, I have real concerns about the contents and concepts of the road map. There's a noticeable discrepancy between the vision for the Middle East that President Bush has expressed in his speeches and that laid out, by implication, in the road map.

In all his public utterances, President Bush has acknowledged that the sufferings of the Palestinian people are not the product of Israeli occupation but of betrayals by the Palestinian leadership. History bears this out. The Palestinians could have had a state at the time of the partition of Palestine and the foundation of Israel, in 1948, when there were no refugees. They could have had a state in 1967, when there were no settlements. They could have had a state in 2000, when there was no *intifada*. In all these critical instances, the Palestinian leadership rejected the option of living peacefully alongside a secure Israel.

In truth, the Palestinian leaders have been more interested in destroying the Jewish state than in building a Palestinian one. That's why the many millions of dollars given to the Palestinian

Authority by the European Union in recent years have gone not to build roads, schools, hospitals, or water treatment facilities for the Palestinian people but to finance terror bombings and to enrich the personal coffers of the Palestinian leaders.

The fact is that Palestinian violence can't be defended as the result of the occupation of Palestinian lands by Israel. It predates that occupation, and it has continued even through periods when Israel has offered extraordinary territorial concessions in an effort to negotiate peace. Now Israel can respond to any new Palestinian peace initiative only after the Palestinians demonstrate they have finally changed, through new leaders, reformed institutions, and the cessation of terrorism. All of these realities underlie President Bush's vision for the Middle East.

The Quartet's road map, on the other hand, starts from a different set of assumptions. The road map assumes that the core of the problem is Israel's occupation of the territories. Time and again, the main goal stated is to end the occupation. Once Israel's occupation is seen as the heart of the problem, what follows in the road map flows inevitably, though disturbingly. Instead of requiring that the Palestinians confiscate illegal weapons and consolidate their security forces in order to achieve a state, the road map merely calls on the Palestinians to commence such a process. On the need to change Palestinian leaders, which President Bush has stressed, there is only a reference to Palestinian leaders who will fight terrorism, leaving open the possibility that today's corrupt leaders could remain in power.

Unlike the Bush vision, the road map sees the Quartet, not the parties themselves, as the ultimate decision makers, something that Palestinians have long desired but that Israel rejects. And the document focuses on the Palestinian strategic goal of a "sovereign, independent, democratic, and viable Palestine" without speaking to Israel's need for an unambiguous renunciation of violence by the Palestinians.

In other words, the road map demonstrates the danger of the international community's becoming the focal point of Middle East diplomacy. According to the road map's logic, Israel is the problem. The need for the Palestinians to truly give up on terror and violence is treated vaguely. If the plan were implemented without any adjustments, Israel would find itself under pressure to make concessions without significant evidence that the Palestinians had changed, that violence and rejectionism had finally come to an end.

In the final analysis, movement by both parties to the conflict will be necessary to create a just and lasting peace. As Sharon now has acknowledged, some of the existing Israeli settlements will eventually have to be dismantled. (It won't be the first time Israel has made such a concession. At the time of the Egyptian-Israeli peace accord, Prime Minister Begin, like Sharon a leader of the "hard-line" Likud Party, dismantled Israeli settlements in the Sinai.) A tough, respected, and realistic Ariel Sharon may well prove to be the ideal statesman to carry out such changes, since all of Sharon's fellow citizens know he would never take a step that would endanger the survival of Israel. Perhaps, as Sharon hopes, it will be possibly to modify the road map in such as way as to transform it into a realistic basis for constructive talks and the first steps toward peace.

Israel has always stood ready to meet the Palestinians halfway on behalf of peace. One thing only has been lacking: a Palestinian partner truly interested in renouncing violence and hatred and ready to help create the conditions under which both Israel and a Palestinian state can live in peace and security. I hope and pray that Mahmoud Abbas may prove to be such a partner.

MEANWHILE, THE SUPPORTERS OF the Palestinian cause throughout the Arab world could play a positive role in creating a partnership for peace if they would renounce their anti-Semitic attacks on Israel and the Jewish people.

It's hard enough to overcome the political and nationalistic problems that stand in the way of Palestinian-Israeli peace. The continuing anti-Semitic incitements in Arab mosques, schools, and media—blood libel charges, conspiracy theories, comparisons of Israel to Nazi Germany—can only embitter the peoples on both sides and make good-faith peace negotiations an even more distant dream.

If continued over time, this barrage of hatred will poison the minds of many in the Arab world so that anti-Semitism will indeed become a permanent way of life. The prospects this development would hold for greater violence against Jews, not only in the Middle East but also around the world, are too terrible to contemplate.

If we continue to let Arab anti-Semitism flourish, it could become one of the most destructive forces unleashed in this new century. History has shown us where anti-Semitism can lead. Combating it right now must be the task not only of Israel and organizations like ADL. For the sake of peace and a sane world, responsible people and governments everywhere must speak up— and do it now.

CHAPTER 8

The Poisoned Well:

Spreading Bigotry Through Popular Culture

IN THE ANNALS of contemporary anti-Semitism, there are some truly hateful and repellent figures—the William Pierces and Matt Hales of the world. People such as these consciously spread hatred and bigotry as widely as possible.

And then there's the case of Dolly Parton.

Dolly Parton? Is this a joke? Does one of America's favorite country singers and entertainers really deserve to be mentioned in the same breath as someone like David Duke?

No, she doesn't. And yet, in my role as director of ADL, I've had run-ins with any number of widely admired, seemingly harmless, and often well intentioned figures from the worlds of entertainment, sports, and politics. The well-known names include those of comedian Jackie Mason, actors Whoopi Goldberg and Marlon Brando, and, yes, Dolly Parton. None of these people, I must stress, is a racist or anti-Semite. All have done good works both as public figures and as private individuals. Yet all have been guilty, on occasion, of helping to spread the virus of bigotry through thoughtless or carelessly malicious words and actions.

In responding to such statements, we must follow principles that are reasonable and fair. We don't want to be guilty of intolerance ourselves. Neither do we want to get caught up in a game of "gotcha!" with public figures. It would harm the cause of freedom and tolerance if we treated people who are guilty merely of foolishness or thoughtlessness the same way we treat the true hatemongers—the neo-Nazis, the Holocaust deniers, the bombers of clinics. So I've given a great deal of thought to drawing appropriate distinctions among various degrees and levels of anti-Semitic speech and action, and responding to people accordingly.

THE ANTI-SEMITIC SPECTRUM

Let me offer some examples that illustrate the spectrum from simple insensitivity to real anti-Semitism. The way these cases were handled illustrates the challenge we face in drawing appropriate moral lines, erring on the side neither of alarmism nor complacency. All happen to involve people from the world of politics.

One case involved a Republican party operative named Fred Malek, who was chairman of the presidential campaign of George H. W. Bush in 1988. When Bush was elected, Malek was named to be White House chief of staff. Then an old story from the years of the Nixon administration surfaced in the *Washington Post.*

Nixon, who was a true anti-Semite, as the most recently revealed White House tapes make shockingly clear, asked his aides John Ehrlichman and H. R. Haldeman to identify high-ranking Jews in the federal government so that steps could be taken to reduce their influence. Fred Malek was a research assistant in the White House at the time, and Ehrlichman and Haldeman told him to prepare the list of Jews. Twice Malek refused. The third time Ehrlichman and Haldeman repeated the request, he did it.

When these facts were revealed a furor erupted, and Malek was widely accused of being anti-Semitic. As a result, Bush had to withdraw his name from consideration for the White House post.

I was one of the few Jewish spokespeople to defend Fred Malek. Look at the context of Malek's actions, I said. Here was a young man working in the White House, under extreme pressure to do the wrong thing. Eventually, he caved in to that pressure. One such mistake, I argued, does not an anti-Semite make. (Unfortunately, I slipped up in my choice of language. In one interview, I said that Malek "was just following orders"—a highly unfortunate phrase to use in this context. But my point stands.) To this day, I am convinced that I was right about Fred Malek.

I later defended President Bush (the elder) against similar charges. Exasperated over lack of progress in the Middle East, Bush complained at a press conference about the undue influence of powerful Jewish lobbyists over Congress. Some called him an anti-Semite for these remarks. I agree that Bush's comments were ill considered. They echoed and thus fed into the classic anti-Semitic fantasy that Jews exert secret control over world government. But considering Bush's overall record, including his long-term efforts to save Jews in the Soviet Union and Ethiopia, I felt he was simply guilty of an ill-tempered remark—and I said so at a press conference of my own.

As these incidents illustrate, it's important for me to stand up to defend public figures when unwarranted charges of anti-Semitism are leveled against them. All I have to offer—as in individual and as a representative of ADL and the Jewish community—is my credibility. If I were reckless about accusations of anti-Semitism, I would quickly lose that credibility and therefore any effectiveness as a leader on this issue.

As it is, when I do accuse someone of anti-Semitic attitudes, my words are taken seriously, because I take them seriously, and people know that.

Drawing the right moral lines can be particularly challenging when the allegations are more serious. Take the case of Joerg Heider, a right-wing politician in Austria whose rise in popularity and influence is a genuine cause for concern among Jews and many others throughout Europe. Heider leads Austria's Freedom Party (FPO), which has steadily gained support in recent elections. In October 1999 FPO drew fully 27 percent of the national vote and earned fifty-three seats in Parliament. Four months later the party was included for the first time in the governing coalition headed by the Social Democrats.

Heider's policies and those of his party are xenophobic and reactionary, centering on imposing restrictions on immigration and cracking down on crime. Heider is also fond of making remarks that minimize the evils of Austria's Nazi past. For example, in July 1991 he praised the Third Reich for the "orderly employment policy" it conducted (ignoring the fact that the Nazi idea of "orderly employment" involved slave labor and concentration camps). And in May 1995 after attending a ceremony in honor of veterans who served in the Waffen SS (Nazi troops personally devoted to Hitler), he defended his participation with the comment, "The Waffen SS was a part of the Werhmacht, and hence it deserves all the honor and respect of the army in public life."

Is Heider an out-and-out anti-Semite? It's a close call. Many Jewish groups think so. He is certainly an unpleasant character and one who is definitely guilty of supporting anti-Semitic causes.

And some of his followers in the FPO cross over the line into undeniable anti-Semitism.

On the other hand, a rabid anti-Semite probably wouldn't visit the Holocaust Museum in Washington, D.C., during a visit to the United States and then issue a statement about the need for universal tolerance and brotherhood, as Heider did.

Our position is to warn against Heider's influence and to closely monitor him and his party. But we refrained from joining other Jewish organizations in denouncing him categorically as an anti-Semite.

More significantly, we also opposed calls for international sanctions against Austria in retaliation for Heider's growing political influence. In my judgment, it would be counterproductive to blame all Austrians for Heider; after all, 73 percent of the voters did *not* vote for his party. Sanctions would isolate Austria and punish the good people of the country for things they're not responsible for.

The ADL was in a minority of minorities on this issue. The result was an uncomfortable time for us in the Jewish community. We were criticized by many in the Jewish press, and Israel canceled its missions to Austria and its student exchanges. I think that was a big mistake—destroying an opportunity to build bridges. The only time I would place an entire country on my "excommunication" list is when the governmental apparatus is turned over completely to the forces of hatred. That happened in Austria in the middle of the last century, but today's situation is quite different. My advice would be: ostracize Heider, yes, but not Austria.

ANTI-SEMITISM AND HOLLYWOOD

All of which brings us back to Dolly Parton.

A perennial fixture on the *Forbes* annual list of America's most successful entertainers, Dolly Parton has a list of accomplishments as long as your arm, ranging from hit songs and popular movies to her own theme park, Dollywood. But when, in 1994, an interviewer from *Vogue* magazine asked her to describe a personal failure, she mentioned her disappointment over being unable to develop a TV miniseries about a born-again Christian folksinger.

And why was this impossible? Parton's explanation: "Everybody's afraid to touch anything that's religious because most of the people out here [in Hollywood] are Jewish, and it's a frightening thing for them to promote Christianity."

When I read this comment, my reaction was one of deep frustration. It's so disappointing to find the old canard about Jewish control of the entertainment industry resurfacing yet again—especially on the lips of someone whose opinions carry weight among country music fans in America's Southern and Midwestern heartland.

It's undoubtedly true that there are prominent Jews among the producers, directors, studio executives, and stars in Hollywood. (Of course, there are also prominent entertainment figures of Italian, Irish, German, African, Hispanic, and Asian descent, among others.) It's even true that, proportionately, there has always been a relatively prominent Jewish presence in the movie, TV, and record industries, for a variety of social and cultural reasons.

Critic Neal Gabler wrote a fascinating history of the subject, *An Empire of Their Own: How the Jews Invented Hollywood,* which chronicles how and why Jewish Americans like Louis B. Mayer, Carl Laemmle, and the Warner brothers came to play such important roles in shaping the movie business. (Anti-Semites sometimes use Gabler's book as evidence supporting their attacks on Hollywood: "You see?" they say. "Even Neal Gabler—a Jew himself—has admitted that the Jews run the industry.")

Does the demographic reality described by Gabler amount to "Jewish control" of the entertainment industry? Not at all. The Jews who work in Hollywood are there not *as Jews* but as actors, directors, writers, business executives, or what have you, just as Dolly Parton is in Hollywood as a singer, actress, and entrepreneur, not as a representative of her favorite church.

Neither does the "Jewish" entertainment industry produce movies or TV programs that somehow promote specifically Jewish values or culture. It couldn't do this even if the Jewish Americans in the industry wanted to because shows that depart drastically from mainstream values simply don't sell—and no one has ever accused Hollywood of being unconcerned with the bottom line. (This explains the paradox that no anti-Semitic conspiracy theorist has ever tackled—how it is that the supposedly Jewish-controlled movie industry has produced so few films dealing with overtly Jewish characters or themes.)

As Neal Gabler explains, the Jewish movie moguls of the 1930s and 1940s actually devoted their careers to lionizing traditional American institutions and values—family, democracy, and free enterprise. They may have been Jews, but on the job they were Americans and entertainers above all; and the same basic pattern persists in Hollywood to this day.

Dolly Parton's ill-conceived comment in *Vogue* ignored these truths and resurrected an old anti-Semitic stereotype that bigots use to attack and discredit not only the entertainment business but the news media, publishing, and other vital links in the world's information network. When the notorious senator Joseph McCarthy launched his attacks on Hollywood in the 1950s, he exploited the same stereotypes, using rhetoric that subtly evoked the anti-Semitic equation "Jew equals Communist" and then tarring an entire industry with the "Red sympathizer" brush. The era of blacklisting and the destruction of many careers ensued.

I'm happy to report that when we publicly challenged Parton's statements (and privately invited her to reconsider), she promptly did so. "I know from personal experience how stereotypes can hurt," she wrote me, "and I regret that my words could have conjured up an impression of Jewish 'control' of Hollywood." In the years since this incident, Parton hasn't reverted to any anti-Semitic flirtation.

Unfortunately, Parton isn't the only Hollywood insider who has thoughtlessly blamed "the Jews" for perceived evils in the movie business. Marlon Brando, one of America's great actors, has often decried Hollywood's use of ethnic stereotypes in pictures, especially in its portrayal of Native Americans in the classic tradition of the Western. He's right about the evils of stereotyping—but he went far astray on the *Larry King Live* talk show in April 1996, when he blamed the problem on Jews. "Hollywood is run by Jews, it is owned by Jews," Brando claimed. He then went on to say that, although movies often depicted ethnic stereotypes of various kinds, "we never saw the kike because they know perfectly well that's where you draw the wagons around."

Of course, ADL promptly denounced Brando's remarks. I'd like to say that Brando responded as quickly and thoughtfully as Parton. Unfortunately, he never retracted his slur. Maybe the silver lining here is that Brando has become an increasingly reclusive, eccentric, and remote figure in recent years. Hopefully most of those who've admired Brando's acting talent recognize that his comments about Jews reflect his own erratic thought processes rather than reality.

It's important that spokespeople like ADL challenge traditional stereotypes like those Parton and Brando revived. If we fail to do so, we seem to tacitly accept them—and to make it easy for some impressionable young person to be receive his or her first dose of the virus of anti-Semitism.

HUMOR AT THE EDGE OF TASTE

There's a school of comedy that deliberately plays at the edges of what's socially acceptable, toying with language, imagery, and ideas that may shock the audience in order to provoke surprise, laughter—and perhaps even fresh insight. It's a tradition that goes back to the satires of Jonathan Swift and includes such popular figures as Lenny Bruce, George Carlin, and Richard Pryor.

Because this style of humor is designed to be intentionally provocative, we need to judge it by different standards than, say, a commencement address, a political speech, or a newspaper editorial. Entertainers certainly deserve the freedom to use distortion, exaggeration, and caricature in the service of humor. And ADL would run the risk of appearing like a bunch of humorless stuffed shirts if we reacted with shock and horror every time a comedian such as Chris Rock or Bill Maher pushed the envelope.

But there are times when I feel I have to speak up, particularly when a humorist steps out of the role of entertainer and speaks as a commentator or ordinary citizen.

Comic Jackie Mason is famous for his stage routines in which he evokes a variety of ethnic and racial stereotypes for shock and comic effect. It's not my favorite kind of humor, but some people like it, and my attitude toward it is one of tolerance. (Although I do feel a little queasy when I hear Mason's fans make comments like, "His routines are so funny—and you know, they have so much *truth* in them, too!")

But I drew the line when Mason appeared as a guest on Pat Buchanan's nationally syndicated radio show in March 2000 and spouted many of the same ethnic stereotypes. Among other unfunny and uninformed statements, Mason claimed that Blacks are naturally inclined toward crime and dismissed the problem of Black-on-Black crime (which statistics prove is by far the most common type of crime committed by African Americans) as "a fallacy." More recently, Mason has taken to repeating equally vicious stereotypes about Arabs, speaking as if every Arab is violent and a terrorist.

These kinds of thoughtless, ignorant statements are deplorable.

Mason likes to claim the mantle of the humorist as a teller of unpopular truths, saying that he's fighting the scourge of political correctness. He's wrong. You don't counter political correctness by spreading hatred and inflicting pain. Mason's forays into amateur political analysis only succeed in encouraging bigotry—and making a fool of himself.

In a similar vein, I think actress and comedian Whoopi Goldberg crossed a line when she contributed a tacky, dumb "recipe" for "Jewish American Princess Fried Chicken" to a fundraising cookbook for the Connecticut community where she lives. The recipe's directions included such witless lines as "Send a chauffeur to your favorite butcher shop for the chicken," "Watch your nails," and "Have Cook prepare rest of meal while you touch up your makeup."

I understand that making fun of human frailty is part of the comic's stock in trade. So if a Henny Youngman wants to tell jokes about his wife's bad cooking while a Roseanne Barr makes fun of some men's addiction to beer and TV sports, so be it. But why label one set of stereotypical traits as "Jewish"? And even more to the point, why publish a would-be satirical recipe in a cookbook where all the other contents are straight, serious recipes for real use in the kitchen? Whoopi's ill-considered attempt at humor was a thoughtless mistake.

HUMOR AND THE HOLOCAUST

Any actor, director, or writer will tell you that humor is perhaps the most delicate, difficult effect to achieve. Treating a serious issue with humor is even more challenging. The danger of a mistake in tone is ever-present. Thus, when I heard in 1999 that Italian comedian and filmmaker Roberto Benigni had created a *humorous* film about the Holocaust, titled *Life Is Beautiful,* I was sure he'd made a big mistake.

Much to my surprise and delight, when I saw the movie I realized that Benigni had created a very special and important work of art.

Life Is Beautiful tells the story of Guido, a hapless, Chaplinesque "little man" caught up in the horrors of the World War II. His humor makes him an endearing soul, and through his comic antics he wins the woman he loves, opens the bookstore he has always dreamed of owning, and settles down to a happy, ordinary life with Dora and their son, Joshua. Life is beautiful—until Guido and Joshua, who are Jewish, are rounded up by the Nazis and sent to a concentration camp.

How Guido saves his young son from death and protects him until liberation is what makes the film unique—and artistically risky. For Guido falls back on what he does best—he activates his comic imagination, convincing Joshua that the entire nightmare in which they are trapped is a game and that Joshua can win the game if he follows the rules Guido sets forth.

The film succeeds by using comedy to create a stark contrast between normal life and life during the Shoah. The humor heightens our anxiety about the terror we know will come, humanizing the tragedy in ways that a more direct depiction might not have done.

The Holocaust is now separated from us by almost sixty years. As time passes and as the number of survivors and witnesses dwindles, we are faced with the challenge of how to impart the lessons of the Holocaust to a new generation. For some high school students today, World War II seems as remote as the Seven Years' War. When they go online to research history, they are bombarded by misinformation alongside fact. Holocaust deniers would have them believe that there were no concentration camps, no Final Solution, no slaughter of the innocents.

At the same time, popular culture trivializes Holocaust imagery. A rude shopkeeper is called a "soup Nazi"; women's rights advocates are labeled "feminazis"; fashion designers create collections deliberately evoking the cut and color of Nazi uniforms; a rap group releases an album entitled *Da Holocaust.* There's a real danger that the awful reality of the Shoah will be forgotten or diminished in the minds of millions.

Life Is Beautiful may pave the way for a new generation to keep the memory alive. By focusing on the personal story of one family, we see at one and the same time how similar to us they are—and yet completely different in the terrible fate they struggled against. It deserves the many honors it has received, including the Grand Jury prize at the 1998 Cannes Film Festival and three Academy Awards. *Life Is Beautiful* joins a small list of films, including *Schindler's List, The Diary of Anne Frank, The Garden of the Finzi-Continis, Shoah,* and Roman Polanski's recent *The Pianist,* which have captured the reality of the Holocaust in the universal language of cinema.

THE DARK SIDE OF THE INTERNET

Movies and music are powerful media for spreading attitudes and images both good and bad. But today no communications medium has greater potential than the Internet. With hundreds of millions of computers already linked through the World Wide Web, countless new communities of interest have emerged, enabling people everywhere to share information, ideas, images, texts, and sound and video files at the click of a key.

Organizations like ADL have directly benefited from the power of the Internet. Through our Web site, we can communicate with friends and supporters around the world more quickly and easily than ever as well as provide vast information resources to students and other individuals interested in the battle against anti-Semitism and other forms of hatred and bigotry.

Unfortunately, the Internet also has a dark underbelly. It's both jarring and profoundly upsetting to go online and see graphic examples of how hatred has migrated from leaflets in parking lots to Web sites and chat rooms. The same extremists who once had to stand on street corners to spew their venom, usually reaching only a few passersby, now can rant from the safety of their own homes with a potential audience of millions—and all while protecting their own anonymity if they wish.

Over the past ten years, in fact, the Internet has enabled those on the far right to create an electronic community of hate that encompasses hundreds of groups and thousands of individuals on every continent. Even worse, it lets this community of hate set snares for unsuspecting children and others who unwittingly stumble into their innocent-seeming Web sites, where the virus of anti-Semitism is ready to be spread.

The story of bigotry on the Internet can be traced back at least to 1995, when Don Black, a former leader of the Knights of the Ku Klux Klan (KKK), established *Stormfront*, the first white supremacist site on the World Wide Web. Since then, as accessing the Internet and creating Web pages has become less expensive and less technologically demanding, hundreds of bigoted sites promoting a variety of hate-filled philosophies have sprouted on the Web.

And because the Internet knows no boundaries, Web bigotry is a global phenomenon. As I'll explain, the most hateful and violence-prone organizations of Islamic extremism have also established footholds on the Web, using the Internet to propagandize, raise money, and (most chillingly of all) to actively support terrorism.

THE EXTREME RIGHT ON THE INTERNET

Let's look first at Nazis in cyberspace. Many groups and individuals have created and maintained Web sites promoting Nazi-style anti-Semitism and racism. We've already discussed the National Alliance, the largest Hitlerian organization in the United States today, whose Web site features transcripts of former leader William Pierce's anti-Semitic radio broadcasts, articles from the group's *National Vanguard* magazine, and a catalog of over six hundred books.

Similar in tone is the Web site of the NSDAP/AO, which is the German acronym for National Socialist German Workers Party—Overseas Organization. The site blames Jews for inflation, media "brainwashing," and governmental corruption while depicting Blacks as criminals and rioters.

Then there's the Web site titled *This Time the World,* which prominently displays a giant swastika, contains speeches by Joseph Goebbels and American neo-Nazi George Lincoln Rockwell, and offers a gallery of Nazi art. Additionally, many neo-Nazi skinheads (violent, racist, shaven-headed youths) such as the Oi! Boys and Hammer Skin Nation have established Web sites, many devoted to racist hard rock music.

Second, we look at the Klan on the Web. Today's Ku Klux Klan is more fragmented than at any time since World War II, but the group's many factions have been using the World Wide Web as a means to revitalization. Spreading the Klan's traditional message of hatred for Blacks, Jews, and immigrants, many Klans and their local chapters are drawing attention to themselves by establishing Web sites.

One Klan site proclaims goals such as maintaining and defending "the superiority of the White race," observing "a marked difference between the White and Negro race," and educating "against miscegenation of the races." Another site claims that Jews killed Jesus and describes them as Satan's people. A third pledges to "stop the uncontrolled, outrageous, and unprecedented plague of immigration."

Among those Klan groups with active Web sites are many chapters of the Knights of the White Kamellia, the two factions of the Knights of the Ku Klux Klan and their subsidiary chapters, the North Georgia White Knights and the Southern Cross Militant Knights. Additionally, the number of Web sites for the National Association for the Advancement of White People (NAAWP), a group founded by former Klan leader David Duke and often described as a "Klan without robes," has grown dramatically in recent years.

A number of sites, including the original *Stormfront,* promote white supremacy in general or espouse some amalgam of the philosophies detailed above. David Duke, America's best-known and most politically active racist, now spreads his slick white supremacist bile on the Internet. Concerned that the "nonwhite birthrate," "massive immigration," and "racial intermarriage" will "reduce the founding people of America into a minority," Duke boasts about the "genetic potential" of "our people," pointing out "innate intellectual and psychological differences" between whites and minorities.

The site for White Aryan Resistance (WAR), a group led by San Diego–based white supremacist Tom Metzger, features unbearably crude caricatures of Blacks and Mexicans while applauding "racial and cultural separatism worldwide." Calling whites "Nature's finest handiwork," Metzger declares, "Your race and only your race must be your religion."

Fellow Californian Alex Curtis, creator of the Nationalist Observer Web site, attacks Jews, Blacks, and immigrants, urging cooperation between "White nationalists, White separatists, Skinheads, National Socialists, Ku Klux Klansmen, and Identity Christians." His "Tribute to Jewry" Web page consists of a picture of "Jew York City" decimated by an atomic bomb. Other sites in this category include *14 Word Press, White Power World-Wide,* and the *Occidental Pan-Aryan Crusader.*

Religious hatred also finds ample expression on the Web. A diabolical mixture of racism, anti-Semitism, and religion, the Identity Church movement uses the Internet to teach that Anglo-Saxons are the Jews described in the Bible, that Jews are the descendants of Satan, and that Blacks and other minorities are inferior "mud people." The *G.O.A.L. Reference Library* Web site contains documents such as *The Talmud: Judaism's Holiest Book Exposed,* which claims that the Talmud preaches violence against Christians, and *Facts Are Facts,* which falsely asserts that today's Jews are not descended from the Jews described in the Bible.

Expressing approval for white separatism, the Web site for militant Identity group Aryan Nations calls Jews "the natural enemy of our Aryan (White) Race," a "destroying virus that attacks our racial body to destroy our Aryan culture and the purity of our Race."

Voicing support for alleged abortion clinic bomber Eric Robert Rudolph, James Wickstrom, and August Kreis, leaders of the violent antigovernment group Posse Comitatus, claim at their Web site that the "federal government has grossly overstepped its bounds" because it is run by Jews, "Satan's kids."

In addition to these Identity sites, *Be Wise as Serpents, Kingdom Identity Ministries, The Lords Work,* and many more are currently available online.

Sharing Identity Christianity's view that nonwhites are subhuman "mud people," the World Church of the Creator (WCOTC) attacks Christianity, Judaism, Blacks, and immigrants with equal vehemence. The group's main Web site blames the Jews for the trade in Black slaves and accuses them of manipulating the government while declaring Blacks physiologically inferior and inherently criminal. WCOTC also presents a number of other attractive, well-designed sites, many of which are adorned by vicious drawings of the group's supporters brutalizing Jews and Blacks.

LURING THE UNSUSPECTING

Many hate sites are being specifically designed to ensnare children and others who may not suspect or understand the real nature of what they are reading. For example, the World Church of the Creator's kids' Web site uses enticing graphics to lure the young and offers simplified versions of WCOTC documents, making them easier for children to understand, along with crossword puzzles and games. A closer look reveals that the games are laced with racist and anti-Semitic themes. The site pictures an idealized portrait of a white family next to the phrase, "The purpose of making this page is to help the younger members of the White Race understand our fight."

The WCOTC *Women's Frontier* Web site presents bigotry behind a veneer of feminism, declaring that the "White female voice must be heard" if the Church is to "truly accomplish its goal of taking back White territory worldwide."

Other sites disguise themselves as legitimate sources of information. There's one site that appears to be an examination of the life of the civil rights leader Dr. Martin Luther King Jr. Any student doing research on Dr. King who might happen upon this site could be duped into believing this is a legitimate history Web site. Among other "educational" features, it includes a "Martin Luther King Pop Quiz" with questions like these:

Question: According to whose 1989 biography did King spend his last morning on earth physically beating a woman?
Answer: Reverend Ralph Abernathy. *And the Walls Came Tumbling Down*

Question: Who was the assistant director of the FBI who wrote a letter to Senator John P. East (R-NC) describing King's conduct of "orgiastic and adulterous escapades, some of which indicated that King could be bestial in his sexual abuse of women"?
Answer: Charles D. Brennan

It takes a trained observer to realize that the site is actually a trove of racist propaganda from the National Alliance.

How the Internet Fosters Hate Crime

If the Internet simply provided hate groups with an alternative means of disseminating their repugnant views, that would be bad enough. But in fact the Internet also provides specific, concrete, measurable support to such groups as they pursue their mission of sowing hatred, discord, and violence. There are three important, measurable respects in which the electronic community of hate strengthens the offline activities work of right-wing extremists:

- The Internet provides *instant and anonymous access to propaganda that inspires and guides criminal activity.*

- It helps hate groups *more effectively coordinate their activities.*

- It offers hate groups *new ways to make money.*

We'll begin by examining how the Internet is being used to encourage hate-based crimes.

Many far-right propagandists encourage their readers to become "lone wolves" (extremists who commit violent crimes alone), and those with the greatest potential to become lone wolves may find the Internet particularly appealing. Beyond finding inspiration online, extremist criminals have found nuts-and-bolts tactical guidance on the Internet.

This situation is markedly different from—and far more dangerous than—the situation before the Internet. For example, consider this incident: in 1988, years before the Internet was widely used, racist skinheads from the group East Side White Pride in Portland, Oregon, attacked three Ethiopian immigrants with a baseball bat and steel-toed boots, killing twenty-seven-year-old Mulugeta Seraw. An investigation of the murder, which resulted in three convictions, revealed intimate ties between the culprits and WAR, the violent white supremacist group led by Tom Metzger.

The national vice president of WAR trained East Side White Pride members in how to attack minorities. As he later explained, "Tom Metzger said the only way to get respect from skinheads is to teach them how to commit violence against Blacks, against Jews, Hispanics, any minority." When the Southern Poverty Law Center (SPLC) and ADL sued Metzger and WAR for their role in the murder of Mulugeta Seraw, the jury awarded $12 million in damages to Seraw's family.

It was a victory for decency, one that we'd hoped would serve as a powerful deterrent to future hate-mongers. But the Internet is tailor-made for extremists wishing to avoid similar lawsuits. Tom Metzger established his Web site in November 1995. On the Internet, Metzger makes his angry rhetoric available to millions without knowing anything about the people who are reading it. Anyone can anonymously visit his site, study his propaganda— and act on it. In fact, Metzger encourages readers to become lone wolves, telling no one of their plans and involving no accomplices who may later testify against them. Thus, for propagandists like Metzger, the Internet is an excellent tool for encouraging violence without paying the consequences.

By the same token, those with the greatest potential to become lone wolves may also find the Internet particularly appealing. While paranoid individuals might refuse to meet in person with others who share their beliefs, they may very well be comfortable reading incendiary propaganda on the Internet, remaining isolated until they violently act out what they have read.

Right-wing extremists can also find plenty of nuts-and-bolts tactical guidance for planning crimes on the Internet. With a click of the mouse, extremist readers can learn about marketing scams, discover how to use "paper terrorism" techniques such as filing specious liens to harass opponents, and master the technologies of violence. Instructions for making bombs and other terrorist tools are readily available online to all types of extremists, and many white supremacist Web sites have either posted bomb-making instructions or provided instant links to such material.

In 1999 British neo-Nazi David Copeland planted nail bombs in a Black neighborhood, an Indian area, and a gay pub in London, killing three and injuring more than a hundred. Copeland later wrote, "I bombed the blacks, Paki's, [and] Degenerates," and he boasted, "I would of bombed the Jews as well if I got a chance." (The errors in grammar are Copeland's, of course.) A court handed Copeland six life sentences for his crimes. He had learned how to build his bombs in a cybercafé, where he downloaded copies of *The Terrorist Handbook* and *How to Make Bombs: Book Two* from the Internet.

"Explosives are very effective in our cause," writes "Death Dealer," the anonymous creator of the racist skinhead site *Better than Auschwitz*. "They should be deployed more." *Better than Auschwitz* includes pictures of bombing victims and detailed bomb-making instructions. The site also features instructions for using knives and brass knuckles in fights against minorities, as well as tips for hand-to-hand combat. A "Nigger Baiting Made Easy" section describes "the various methods of selecting muds and queers, and getting them to fight, or throw the first punch." Such material resembles the instructions White Aryan Resistance gave the skinheads of East Side White Pride before their violent rampage in Portland.

Extremists can also find online guidance about *whom* to attack. Antigovernment sites frequently post information about judges, law enforcement officers, government officials, and other potential victims of violence. For example, the notorious *Nuremberg Files* Web site developed by violent antiabortion extremists presented graphics dripping with blood and links to sites calling the murder of abortion providers "justifiable." Nearby was detailed personal information about doctors who allegedly provide abortions, including their social security numbers, license plate numbers, and home addresses.

The list of doctors reads like a list of targets for assassination. Names listed in plain black lettering were "still working"; those printed in grayed-out letters were "wounded"; and those names that were crossed out indicated doctors who have been murdered ("fatality"). At the site, the name of Dr. Barnett Slepian, who was murdered in his upstate New York home by a sniper in 1998, was crossed out within hours of his death. After a 1999 court ruling held antiabortion groups liable for $100 million in damages for providing information to the site, it was temporarily redesigned to eliminate the crossing out of murdered doctors' names. However, the site is still up and running, listing not only doctors who provide abortions but also their spouses, judges who have ruled in favor of abortion rights, and others deemed criminal by the site's creators.

WHAT CAN BE DONE?

Everything we've discussed in this chapter raises the question: What should be done about this spreading of hatred through cyberspace?

Most people who come to understand the scope of the problem say, "There ought to be a law." It's an understandable reaction. But the First Amendment of the United States Constitution protects the right of all Americans to express their opinions, even if they make statements that are unpopular or downright offensive. Federal, state, and local governments may intrude upon this right in only very limited situations. In the "marketplace of ideas" model adopted long ago by the Supreme Court, good and bad ideas are free to compete, with truth ultimately prevailing.

Thus Americans are willing to tolerate harmful speech because they believe that in the end it will be tested and rejected. This philosophy applies to the Internet just as it applies to the print and broadcast media or even the traditional soapbox in a public park.

For this reason large portions of the Communications Decency Act (1996), which dealt not only with Internet pornography but also with the hate speech of extremist groups, were ruled unconstitutional by the Supreme Court. Other attempts to regulate the Internet in the United States have been struck down for the same reason.

However, there are legal remedies available when hate speech crosses the line into threats and intimidation. Under the law, threats are not protected under the First Amendment. This applies to threats involving racial or ethnic epithets or those motivated by racial or ethnic animus. And recent decisions, such as *Planned Parenthood of the Columbia/Willamette v. American Coalition of Life Activists,* have found online threatening speech to be punishable. Thus a threatening private message sent over the Internet to a victim, or a public message displayed on a Web site describing intent to commit acts of racially motivated violence, can be prosecuted under the law. Similarly, harassing speech is not constitutionally protected because the speech in question usually amounts to impermissible conduct, not just speech.

Note that, to fall under the jurisdiction of the law, both harassment and threats must be directed at specific individuals or organizations. Blanket statements expressing hatred of an ethnic, racial, or religious group cannot be considered harassment, even if those statements cause emotional distress.

Likewise, the concept of "group libel"—comments directed toward Jews, Blacks, or any other religious or racial group—cannot be used as a weapon against haters who spew invective online or off. The courts have repeatedly held that libel directed against religious or racial groups does not create an actionable offensive. Libel on the Internet directed against a particular person or entity, of course, is actionable under the law just like libelous remarks uttered in any other public forum.

Another unprotected activity is incitement to violence. However, the Supreme Court ruled in the case of *Brandenburg v. Ohio* that there is a line between speech that is "directed to inciting or producing imminent lawless action" and speech that, while hateful, is not likely to incite such action. The *Brandenburg* standard sets the bar quite high. Because it's difficult to prove that a specific Web site led directly and imminently to an act of violence, online hate speech will rarely be punishable under this test.

While hate speech online is not in itself punishable, it may provide evidence of motive in a hate crime case. Forty-two states and the District of Columbia currently have some form of a hate crime law on the books that enables prosecutors to seek increased penalties when a victim is targeted in a bias crime. (Most of these laws are based on model legislation originally drafted by ADL in 1981; a Wisconsin law following the ADL model was upheld by the United States Supreme Court in June 1993.) When hate speech inspires violence, the evidence could aid the prosecution in seeking an increased penalty under the hate crimes statute.

A good example is the landmark *Wisconsin v. Mitchell* case, in which the Supreme Court upheld the application of hate crime law by a 9–0 vote. The defendant incited a group of young Black men who had just finished watching the movie *Mississippi Burning* to assault a young white man by asking, "Do you feel all hyped up to move on some white people?" The court found that the statement was appropriately used by prosecutors to demonstrate that the white victim was assaulted because of his race. It seems logical that hateful views expressed on the Internet could be used in a similar manner.

While this concept has been applied thus far only to movies, there have been an increasing number of crimes being committed by perpetrators who read hate literature online. The racially

motivated shooting of Blacks, Asian Americans, and Jews in suburban Chicago over the Fourth of July weekend in 1999 was carried out by a member of World Church of the Creator, Benjamin
Nathaniel Smith, who, according to law enforcement officials, has
admitted to reading hate literature online. There have been similar cases where perpetrators of hate crimes have found inspiration
in literature easily obtainable on the Internet. In time, such a case
is likely to appear before the Supreme Court, and the ruling that
results will have an important impact on the battle against hate
speech.

Even with laws against intimidating speech, the anonymity of
the Internet makes it difficult to track down and prosecute perpetrators of threatening messages. This proved true in a recent
case involving a Detroit boy who received a barrage of anti-
Semitic death threats in his mailbox. The eleven-year-old, who
innocently stumbled upon a hater while surfing through a public
chat area, immediately reported the incident to his parents, who
notified the local police. Not surprisingly, their investigation
turned up few clues as to the source of the anonymous threats.
Eventually, it was determined that the source was disguised, quite
possibly outside of the country, and obviously well beyond the
reach of local authorities.

Yet there have been other successful prosecutions against
senders of hate mail. A student at the University of California,
Irvine, who transmitted threatening e-mails to sixty Asian students was caught and convicted of a civil rights violation (*United
States v. Machado*, 1998). There have been other similar convictions.

Even more clearly illegal is fund-raising by terrorist groups on
the Internet, which violates the laws of the United States and
those of many other nations. American laws such as the Antiterrorism and Effective Death Penalty Act (1996) and the USA
Patriot Act (2001) prohibit providing material support to groups
identified as terrorist by the State Department. All of the Islamist

groups discussed earlier in this chapter are found on the State Department list of foreign terrorist organizations.

From a commercial standpoint, the willingness of Internet service providers (ISPs) to host Web sites that promote hatred can also create a barrier against online bigotry. However, while many Internet access providers have policies that regulate offensive speech, most do not ban hate speech outright. Some providers cite First Amendment rights as reason enough not to interfere with content on their servers.

ADL has worked closely with several major Internet companies to establish and enforce clear guidelines regulating what is acceptable and unacceptable on their sites. AOL, for example, has withdrawn its hosting service from Web sites (such as one operated by the Texas Ku Klux Klan) that peddle bigotry online. Online auctioneer eBay has a policy that forbids the sale of items "that promote or glorify hatred, violence, or racial intolerance" while establishing carefully defined exceptions for items of historical interest—for example, books and documentary movies about Nazi Germany. And Barnes & Noble prominently posted a factual ADL warning about the contents and provenance of *The Protocols of the Elders of Zion* on the Web page where that book is offered for sale.

Unfortunately, some Internet service providers have been less willing to establish firm policies against hate speech. For example, Earthlink of Pasadena, California, states in its "acceptable use policy" that the site "supports the free flow of information and ideas over the Internet" and does not actively monitor the content of Web sites it hosts. Although Earthlink makes it clear that illegal activities are not permitted on its site, that caveat didn't stop the neo-Nazi Web site *For Folk and Fatherland* from establishing a home page through Earthlink. The Web site reprints (without disclaimer or warning) Hitler's *Mein Kampf* and more than two dozen of Hitler's speeches. It's not illegal activity, but the message is clearly hateful.

Those hate groups that have trouble gaining access to mainstream Internet service providers can turn to one of a number of renegades of the Web, hate institutions such as Don Black's *Stormfront*. Since becoming the first hate site to go live, in 1995, *Stormfront* has leaped into the business of hosting extremist sites, describing itself as "an association of White activists on the Internet whose work is partially supported by providing webhosting for other sites." At least one extremist bumped from a mainstream online service has taken refuge on Black's server. Alex Curtis's Nationalist Observer site, once hosted by America Online, now resides at *Stormfront*. The implication is clear: no matter how many mainstream Internet providers rebuff the bigots, those determined enough to establish a racist site will be able to find a willing host.

Clearly, censorship is not the answer to hate on the Internet. ADL supports the free speech guarantees embodied in the First Amendment of the United States Constitution. The best antidote to hate speech, I've always maintained, is more speech. Public awareness of hate on the Internet, whether through reports and studies or media coverage, can go a long way to help sensitize the public, private Internet companies, and government regulators to the problem.

ADL continuously monitors and documents Internet hate. By communicating its findings, ADL promotes public awareness of the plans and history of online bigots, in line with the league's view that exposure will lead to rejection of haters and their propaganda.

One concrete method for combating hate speech online that is available to individual computer users is to deny the bigots access to home computers. ADL has developed software for this purpose in cooperation with The Learning Company (TLC) of Massachusetts, using the technology of TLC's *CyberPatrol*® software. This software, titled ADL *HateFilter*™, provides parents and others with the ability to block access to Internet sites that ADL believes promote hate directed at groups or individuals that are singled out because of their religion, race, ethnicity, gender, or sexual orientation. The software redirects the computer user to information about hate groups at the ADL home page.

HateFilter does not seek to prohibit hate speech on the Internet—that would probably be impossible as well as unconstitutional. However, it recognizes that the Internet is different from libraries and bookstores. In those locations, questionable material can be labeled and organized in a way that enables parents to exercise discretion about what their children see. *HateFilter* is an attempt to afford parents the ability to exercise similar discretion over the Internet.

There are no simple answers to these problems. Yet we as a society must find a way to respond to the challenge. Internet users need to let responsible authorities know about the threatening, hateful, and violent material they find online. Parents and teachers need to recognize warning signs in the behavior, language, and attitudes picked up by Web-surfing youngsters. And the computer industry, educators, parents, civil rights groups, and government agencies must work together to develop new and creative approaches to the unprecedented challenges posed by online extremism.

A balance between the need to protect our cherished freedoms and the need to discourage the online dissemination of bigotry, violence, and hatred must be found—and I believe it will.

Not So Harmless

Let me return to the point with which this chapter began. There's a world of difference between a Dolly Parton, who repeated a traditional anti-Semitic slur out of ignorance or thoughtlessness, and a Tom Metzger, who deliberately uses the Internet to encourage hate crimes and violence. It's tempting to think we can focus solely on exposing and stopping the death dealers like Metzger and ignore the foolish anti-Semitic words and deeds of public figures (or private acquaintances).

A lifetime of labor on behalf of tolerance has shown me that this is not a viable option. When good people allow public statements that encourage stereotyping, prejudice, and hate to go unchallenged, the bar of civil discourse in our society is subtly lowered. For impressionable young people, a signal is sent: It's okay to make vicious statements about groups you don't like. And those who want to go a step further and actually encourage racist or bigoted actions, up to and including violence, are encouraged. They eagerly interpret our silence in this way: "You see? Deep down, most people *agree* that the Jews [or Blacks or gays or Hispanics] are evil. That's why nobody protests when someone dares to speak the truth."

Furthermore, when prominent figures from popular culture adopt extremist views, they can use their wealth, their influence, and their public platform to spread those views far more widely than any bigot in the street could do. Take actor and director Mel Gibson. He is one of Hollywood's most successful leading men, whose roles in such hit films as *Mad Max,* the *Lethal Weapon* series, and *Braveheart* have won him the admiration of millions of fans around the world.

It so happens that Gibson belongs to a splinter group of Christians who call themselves Catholic traditionalists. Numbering some one hundred thousand adherents in the United States, Catholic traditionalism holds that the current leadership of the Roman Catholic church is illegitimate; that the reforms instituted by the Second Vatican Council were evil; and that the church must renounce its sinful modern ways and return to the faith and practice of three centuries ago. Among other things, this requires that worship services be held in Latin rather than in the languages of the people.

So far, the movement may sound harmless, if a bit eccentric. But when you scratch the surface, a more disturbing picture emerges. It turns out that many Catholic traditionalists subscribe to the same kinds of conspiracy theories and anti-Semitic ideas that extreme right-wing groups espouse. In a recent interview an author and activist who is prominent in Catholic traditionalist circles held forth about the evils of Vatican II, which, he darkly intimated, was "a Masonic plot backed by the Jews." The church's enemies, he added, might have threatened "to atom-bomb the Vatican City" as part of a (successful) plot to install their puppet pope, John XXIII. He even dismissed the Holocaust as a phony story concocted by Hitler and unnamed "financiers" as part of a plot to move Jews out of Germany.

Who is this Catholic traditionalist leader with the frightening, extreme views? His name is Hutton Gibson, and he is Mel Gibson's father.

The actor may or may not subscribe to his father's paranoid fantasies; he avoids discussing the details of his faith in public. But in a 1995 *Playboy* interview he dropped some hints about his belief in conspiracy theories that explain much of American history, before changing the subject with the remark, "I'll end up dead if I keep talking."

As the most prominent (and probably the wealthiest) member of the Catholic traditionalist sect in America, Gibson is using his resources to support its growth. He has given over $2.8 million to build a church in the hills near Malibu where a traditionalist congregation gathers. More disturbingly, he has reportedly invested up to $25 million in the development of a film called *The Passion,* now in production in and around Rome, in which Gibson hopes to tell the "true" story of the death of Jesus.

According to a family friend who is also a Catholic traditionalist, "the film will lay the blame for the death of Christ where it belongs." The implication is disturbing. *The Passion* will try to revive the traditional charge of deicide against the Jews—and it will do so armed with the power of a multimillion-dollar Hollywood marketing campaign and the star power of one of the world's best-liked leading men. If the movie becomes a hit, millions of young viewers around the world who innocently assume that *The Passion* is historically and theologically accurate will absorb the lesson that Jews are "Christ killers." It's a belief that has already been responsible for countless deaths.

I hope that, when *The Passion* appears, it will prove to be theologically and socially harmless. If so, I won't hesitate to say so publicly. But I'm very worried that this will not be the case.

The case of Mel Gibson illustrates the kind of damage that can be done when thoughtless or ignorant views like those expressed by Dolly Parton are allowed to take root and grow. That's why such views must be challenged as soon as they appear.

Artists, entertainers, and other public figures don't lose their First Amendment rights when they become famous. Stars have the same freedom as any other citizens to promote their favorite political, social, and religious causes. But the huge megaphone they wield as a result of their fame carries with it a disproportionate degree of influence. When an important figure of pop culture begins to use his or her power in support of bigotry or hatred, I view myself as honor bound to speak up.

In truth, all people of goodwill should view themselves as sharing the same responsibility. When our shared culture is polluted by anti-Semitism, racism, ethnic stereotyping, and other forms of bigotry, we all suffer. Particularly in today's dangerous world, it's important for all of us to draw a line in the sand that defines what is acceptable and what is not, and then to defend that line with clarity, consistency, and courage.

Now More than Ever

ANTI-SEMITISM, APPARENTLY THE OLDEST and most resilient form of hatred known to humankind, has recently moved into an alarming new phase, crossing boundaries of every type—geographic, national, political, religious, and cultural.

A frightening coalition of anti-Jewish sentiment is forming on a global scale. We see it in dozens of manifestations.

We see it in the subversion of the UN-sponsored Durban conference on racism, which turned into a festival of anti-Israel, anti-Zionist, and anti-Jewish hatred.

We see it in the spread of outlandish conspiracy theories, including the newest version of the Big Lie: the bizarre claim that the Israeli Mossad destroyed the World Trade Center on September 11, bolstered by the fabricated "evidence" that four thousand Jews did not report for work that day because they were in on it.

We see it in the scores of racist and anti-Semitic Web sites that pollute the Internet, enabling hate-mongers to cross-pollinate globally and spread their venom on a scale never before possible.

We see it in the Arab mass media, where the proliferation of vicious Nazilike stereotyping of Jews, conspiracy theories, and Holocaust denial messages are poisoning the minds of a generation of Muslim youth.

We see it in the messages of hate preached in Middle Eastern mosques and broadcast electronically around the world, influencing Muslim immigrants in Europe to commit acts of vandalism and violence against Jewish victims.

We see it in the proliferation of outrageous comparisons between fascism and Zionism—the depiction of Israelis as Nazis, of Jewish leaders as Hitlers, and of Israeli treatment of Palestinians as worse than the Holocaust.

We see it in the resurgence of age-old anti-Semitic stereotypes, frauds, and forgeries, including the reappearance of the long-discredited *Protocols of the Elders of Zion* and the continued spread of the infamous "blood libel" that blames Jews for the murder of innocent non-Jews.

Most alarming, these signs of the new anti-Semitism are visible on every continent and in virtually every country of the world.

The last half-decade has witnessed a horrific resurgence of anti-Semitism in Europe, less than sixty years after the murder of six million Jews in the Holocaust—the crime of crimes that many of us believed would make a rebirth of full-blown anti-Semitism practically impossible. Physical attacks on Jews and their institutions are taking place from France to Russia, Spain to Poland, while governments look the other way or respond too slowly and too ineffectually. Leading European news media are filled with stories slanted against Israel, further heating up a climate in which leadership of the Jewish community is virtually alone in its battle against anti-Semitic attacks. In England, Christian clerics are using anti-Jewish rhetoric suppressed since the Holocaust. Not only are we hearing the cry of "Zionism is racism" and attacks on the legitimacy of the state of Israel, but also the old, destructive "replacement theology"—the notion that Judaism has been superseded as a religion—which we thought Christianity had finally outgrown.

In the nations of Asia and Latin America—even in countries where Jews are few in number—anti-Semites are spreading their hatred, often thinly disguised as anti-Zionism. As a consequence, Jewish communities large and small are becoming increasingly vulnerable. Worshipers on the High Holy Days are attacked; synagogues are targeted by rock throwers and bomb planters; children who dare to wear the Star of David are subject to taunts and beatings.

And even in America, the land of freedom and tolerance where Jews historically have enjoyed greater acceptance than anywhere else other than the state of Israel itself, disturbing signs of the spread of anti-Semitism are visible. On college campuses, peace demonstrations all too often degenerate into attacks on Israel, replete with thinly veiled anti-Semitism language and imagery. Among Black Americans, too many leaders find it advantageous to flirt with anti-Semitism in the guise of communal pride. And some politicians of both left and right are beginning to toy with the idea of blaming Israel and "the Jewish lobby" for everything from the war in Iraq to the difficulties of the ongoing battle against terrorism.

Thanks to modern methods of communication, from cable television and cell phones to the Internet, the latest anti-Semitic whispers spread faster, farther, and more quickly than ever. As a result, a worldwide community of hate is developing that links seemingly incompatible forces into a loose network of bigots with many shared enemies, objectives, and policies. Individuals and groups who otherwise agree on little find common cause in their hatred of the Jews. Thus African American anti-Semites are swapping ideas with extremist Islamic clerics from Indonesia and Pakistan; xenophobes and reactionaries from the European far right are abetting the efforts of Holocaust deniers at universities in the United States and Britain; armed nationalist militants in the Rocky Mountains are studying and learning from the acts of terrorists from Saudi Arabia and Palestine.

People are not born bigots. They must be taught to hate. And today millions are learning that deadly lesson. In a time of growing political tensions, cultural anxieties, and economic uncertainties, the age-old temptation to lash out at "the other" is too enticing for many to resist—whether that "other" is Jewish or Black, Arab or Hispanic, immigrant or gay. And when people indulge that temptation to scapegoat and to hate, violence and death result, as history and the pages of today's newspapers make all too clear.

It's time to join forces against this spreading evil before it's too late. Now, more than ever, the motto "Never again!" must become a rallying cry for all people of goodwill—for Jews and Christians, Muslims and Hindus, and lovers of freedom everywhere who profess no creed. Our responsibility is to speak out and to act when tolerance is threatened—and not just in our own backyards or when our own families or friends are in danger, but in all times and places.

I've written *Never Again?* as a wake-up call—a warning about a looming international emergency. By reading this book and absorbing its lessons, you've already taken a first step in combating the spread of anti-Semitism and other forms of bigotry. Here are some other steps we can take—as individuals and as citizens of our nation and the world.

- Demand honesty from local and national authorities, civic organizations, and members of the news media in confronting words and deeds that embody hatred. Call attacks on Jews and Jewish institutions what they are—acts of anti-Semitism.

· Call on political, religious, and civic leaders to use the opportunities they have every day to speak out against bigotry and in favor of tolerance and freedom.

· Urge local officials to use the full resources of the law to combat hate crimes. When acts of violence or harassment are motivated by ethnic, religious, or gender bias, not just an individual but the entire community is under attack. Therefore, tougher punishments are warranted and should be sought.

· Work for the passage of hate crime legislation on the national level and in states and localities where no such laws currently exist.

· Support law enforcement authorities in their response to hate crimes and in their efforts to prevent them. Encourage your community to invest time and money in providing police and emergency workers with the skills and knowledge they need to respond sensitively to incidents of bias or hatred.

· Protest when you see casual anti-Semitism and other forms of bigotry in the media, in the entertainment world, in schools, or in houses of worship. When you encounter bigoted attitudes in private life—in the workplace or in your community—have the courage to speak out for tolerance.

· Encourage schools, libraries, community centers, and other civic organizations to provide education on the dangers of bias and the importance of mutual respect. If your children's schools don't currently offer a tolerance curriculum, urge them to adopt one.

· Teach tolerance at home. Monitor your children's diet of TV, music, movies, books, and the Internet, and make sure they know how to recognize and reject messages of hatred. Even more important, model tolerance and respect in your own life, since no one has a more profound influence on your child's attitudes and actions than you.

· Make a personal effort to develop cross-cultural skills, including the ability to listen to, learn from, and empathize with people from a wide variety of ethnic, religious, racial, and national backgrounds. Attend a worship service with a friend of a different faith from your own, or enjoy a family meal with someone of a different background. Look for the common ground in our shared humanity.

There's no better way for me to close than with the words of Elie Wiesel, the much-admired philosopher, Holocaust survivor, and winner of the Nobel Peace Prize: "Whenever men or women are persecuted because of their race, religion, or political views, that place must—at that moment—become the center of the universe."

All of us who profess to love freedom must learn to live in the spirit of those words, for in a world as deeply and closely linked as ours, only when tolerance is the rule in every land can the freedom of anyone be truly secure.

Never Again? is not intended as a work of scholarship but rather as a personal testament and warning concerning the dangerous revival of anti-Semitism in our time. Therefore, the following source notes should not be regarded as a comprehensive bibliography of important writings concerning anti-Semitism but simply as a list of some of the materials I found helpful while preparing the text. The same materials will likely be useful starting points for readers who may want more detailed information about particular topics discussed in the book. In cases where complete citation information appears in the text, no additional source note is provided.

INTRODUCTION

Page 1 : First two news stories on this list from "Global Anti-Semitism: Selected Incidents Around the World in 2002,"

online at adl.org/Anti_semitism/anti-semitism_global_
incidents_2002.asp. Other news stories on this list from the
database of the Stephen Roth Institute for the Study of
Contemporary Anti-Semitism and Racism at Tel Aviv
University, online at www.tau.ac.il/Anti-Semitism/
database.html.

1. As Storm Clouds Gather

Page 7: "American white supremacist David Duke . . ." From the
database of the Stephen Roth Institute. Other news stories on
this list are from *Antisemitism Worldwide 2000/1,* Anti-Defamation
League and World Jewish Congress, New York, 2002.

Page 8: "Matt Hale, leader of the white supremacist . . ." and
other news stories on this list from "U.S. Anti-Semites Take
Up Palestinian Cause," ADL online at adl.org/extremism/
extr_palestinian.asp.

Page 9: "At protest marches against the 2003 war in Iraq . . ." From
the database of the Stephen Roth Institute. Other news stories
on this list from "Anti-Semitic/Anti-Israel Events on Campus,"
ADL online at adl.org/CAMPUS/campus_incidents.asp.

Page 11: "Monitoring worldwide developments in this deepening
crisis . . ." Details in next six paragraphs from "State of Anti-
Semitism in Four Countries: National Representatives
Report," October 31, 2002, ADL online at adl.org/Anti_
semitism/conference/as_conf_countries.asp.

Page 14: "The following are the eleven statements . . ." Survey
items from *European Attitudes Toward Jews: A Five Country Survey,*"
Anti-Defamation League, New York, 2002.

Page 15: "In 2002, the anti-Semitism index was employed . . ."
Survey results from *European Attitudes Toward Jews: A Five Country
Survey,*" ADL.

Page 18: "In a 1968 speech at Harvard University . . ." From

"Letter to an Anti-Zionist Friend," *Saturday Review* XLVII (August 1967), p. 76.

Page 22: "Now over 15 million Muslims live . . ." From "Europe's Muslim Street" by Omer Taspinar, *Foreign Policy,* March/April 2003.

Page 22: "In the United States, the same trend is apparent . . ." Data in this paragraph from the *American Jewish Year Book,* American Jewish Committee, New York, 2002, and "Fact Sheet: Islam in the United States," U.S. Department of State, International Information Programs, online at usinfo.state.gov/usa/islam/fact2.htm.

Page 23: "For instance, when a French government spokesman . . ." From "A Slander on France" by Francois Bujon de l'Estang, *The Washington Post,* June 22, 2002.

Page 24: "We see similar movements gaining a foothold . . ." Data in next two paragraphs from *Antisemitism Worldwide 2000/1,* Anti-Defamation League and World Jewish Congress, New York, 2002.

Page 26: "In August 2001 Israel appointed a new ambassador to Denmark . . ." From "Israeli ambassador faces torture protest," BBC News, August 15, 2001, online at news.bbc.co.uk/1/hi/world/middle_east/1491978.stm.

Page 26: "Daniel Bernard, the French ambassador to Great Britain . . ." From "'Anti-Semitic' French envoy under fire," BBC News, December 20, 2000, online at news.bbc.co.uk/1/hi/world/europe/1721172.stm.

Page 33: "So Malaysia's Prime Minister, Mahathir Mohamad . . ." From "Malaysia Premier Sees Jews Behind Nation's Money Crisis," by Seth Mydans, *New York Times,* October 16, 1997.

Page 34: "What's more disturbing is when these bizarre lies . . ." From "Somebody Blew Up America: A poem about 9–11 by Amiri Baraka 1oct01," online at mindfully.org/Reform/2002/Amiri-Baraka-Somebody-Blew-Up.htm.

Page 34: "According to a Gallup poll released in March, 2002 . . ." Cited in "Is the Muslim world still in denial about September 11?" by Barbara Amiel, the *Daily Telegraph*, March 4, 2002, online at hvk.org/articles/0502/181.html.

Page 34: "Syria's ambassador to Tehran, Tartky Muhammad Sager . . ." From *MEMRI Special Dispatch Series*, Number 446, December 3, 2002, The Middle East Media Research Institute, online at memri.org/bin/opener.cgo?Page=archives&ID= SP44602.

Page 36: "We charge that a cabal of polemicists . . ." From "Whose War?" by Patrick J. Buchanan, *The American Conservative*, March 24, 2003.

Page 36: "For example, a March 2003 poll by the Pew Research Center . . ." From "Different Faiths, Different Messages: The Issue of War," March 19, 2003, Washington, D.C., The Pew Charitable Trusts.

Page 36: "Democratic Congressman James P. Moran of Virginia declared . . ." From "Virginia Pol 'Regrets' Remarks on Jews," CBSNews, online at cbsnews.com/stories/2003/03/14/ politics/main544089.shtml.

2. JEWISH IN A HOSTILE WORLD

Page 41: Definitions of anti-Semitism from *Living With Antisemitism: Modern Jewish Responses,* edited by Jehuda Reinharz, University Press of New England, Hanover, New Hampshire, 1987, and *The Encyclopedia of the Jewish Religion,* edited by R. J. Zwi Werblowsky and Geoffrey Wigoder, Adama Books, New York, 1986.

Page 43: "Another example is Bobby Fischer . . ." From "The Madness of King Bobby," by Rene Chun, *The Observer,* January 12, 2003, online at observer.guardian.co.uk/osm/story/0,6903, 870785,00.html. Also see "Bobby Fischer's Pathetic Endgame," by Rene Chun, *The Atlantic,* December 2002.

Page 44: Quotation from Book of Esther from the Bible published by the Jewish Publication Society of America, Philadelphia, 1955.

Page 48: "Scholars such as the late Dr. James Parkes . . ." Notable writings by Dr. Parkes include *Judaism and Christianity* (Chicago: University of Chicago Press, 1948), *End of an Exile: Israel, the Jews and the Gentile World* (London: Vallentine, Mitchell, 1954), *Antisemitism* (London: Vallentine, Mitchell, 1963), and *Prelude to Dialogue: Jewish-Christian Relationships* (London: Vallentine, Mitchell, 1969).

Page 49: "Officially, the libel was disavowed by the church . . ." *Constantine's Sword: The Church and the Jews* by James Carroll, Houghton Mifflin, Boston, 2001, pages 272–273.

Page 50: "The first warning signs of this new form of anti-Semitism . . ." *Racism: A Short History* by George M. Frederickson, Princeton University Press, Princeton, NJ, 2002, pages 31–32.

Page 52: "Perhaps, as the American novelist and social critic Mark Twain supposed . . ." *Concerning the Jews,* originally published in *Harper's New Monthly Magazine*, March 1898; reprinted by the Anti-Defamation League, New York, 1992.

Page 53: "Novelist Anne Roiphe speculates . . ." "Antisemitism: Our Constant Companion?" From *Antisemitism in America Today: Outspoken Experts Explode the Myths*, edited by Jerome A. Chanes, Birch Lane Press, New York, 1995, page 459.

Page 57: "Here's an example of what I mean . . ." Career details from "Biography: General George S. Brown," online at www.af.mil.

Page 61: "As columnist George Will has pointed out . . ." From "Final Solution, Phase 2," by George Will, *Washington Post,* May 2, 2002.

Page 68: "Some Jews, especially some of the ultra Orthodox . . ." Some details from "Jews Against Israel," by Michelle Goldberg,

online at salon.com/news/feature/2003/03/13/neturei_
karta/index.html.

3. Cradle of Hatred

Page 76: "And recent scholarship continues to unearth evidence
. . ." *The Popes Against the Jews* by David I. Kertzer, Alfred A.
Knopf, New York, 2001, page 9.

Page 77: "I'm sad to report that Father Gumpel . . ." From
"Vatican official criticizes Jews," by John L. Allen Jr., *National
Catholic Reporter,* December 11, 1998.

Page 95: "There have been other times during the papacy of John
Paul . . ." From "Pope Chided for Silence Amid Slurs on Jews,"
by Joe Mahoney, the *New York Daily News*, May 12, 2001.

Page 95: "After the pope's visit to Israel in 2000 . . ." From "Bad
Diplomacy" by Arthur Hertzberg, online at beliefnet.com/
frameset.asp?boardID=1807&pageloc=/story/17/story_1719_1
.html

Page 98: "Even more disturbing is the role . . ." Details in this list
from *Antisemitism Worldwide 2000/1,* Anti-Defamation League and
World Jewish Congress, New York, 2002.

4. Danger on the Right

Page 102: "One American neo-Nazi said of the September 11
terrorists . . ." From an e-mail sent by Billy Roper to members
of the National Alliance, September 11, 2001.

Page 103: "In fact, in the fall of 1994, some six months prior . . ."
Armed & Dangerous: Militias Take Aim at the Federal Government, New
York, Anti-Defamation League, 1994.

Page 105: "For example, the members of the so-called Phineas
Priesthood . . ." From "The Kids Got in the Way," by Frank
Gibney, Jr., *Time* magazine, August 23, 1999, page 24, and "Va.
Author in Spotlight After Racial Shootings," the *Daily Press,*
Newport News, Virginia, September 2, 1999, page C4.

SOURCE NOTES

Page 115: "One popular item distributed by the NA . . ." From "Don't Think Twice, It's All White," by David Mills, the *Washington Post,* May 16, 1993, page F1, and "'The Saga of White Will' Filled With Racist Trash," by Mike Sangiacomo, the *Cleveland Plain Dealer,* October 31, 1993, page 6H.

Page 115: "Essential to the group's vision is the creation . . ." From "Supremacists Seeking Maine Foothold," by David Connerty Marin, the *Portland* [Maine] *Press Herald,* December 18, 2002, page 1B, and "Fighting Bigotry," by Denise Smith Amos, the *Cincinnati Inquirer,* January 15, 2003, page 2A.

Page 117: "Pierce's follow-up to the *Diaries* . . ." From "New Face of Terror Crimes: 'Lone Wolf' Weaned on Hate," by Jo Thomas, the *New York Times,* August 16, 1999, page A1, and "Lone Wolf Activism," by Katherine Seligman, the *San Francisco Chronicle,* June 4, 2001, page A3. *Hunter* was published by National Vanguard Books, 1989.

Page 119: Some details and quotations from song lyrics on the next several pages are from *Soundtracks to the White Revolution: White Supremacist Assaults on Youth Music Subcultures,* edited by Davin Burghart, Center for New Community, Chicago, 1999.

Page 126: "Both on its national and chapter Web sites . . ." Online at www.cofcc.org.

Page 126: "A November 1999, article in the *Augusta Chronicle* (Georgia) . . ." From the *Augusta* [Georgia] *Chronicle,* November 28, 1999.

Page 127: "In an article posted on the Arkansas Web site . . ." Online at www.arcofcc.freeservers.com.

Page 128: "As Robert Patterson, the publication's past editor, has written . . ." From *The Citizens Informer,* Volume 29, 3rd Quarter, 1998. Online at www.templeofdemocracy.com/Citizens Informer.htm.

Page 129: "However, the ante was upped a few days later . . ." From "Lott Renounces White 'Racialist' Group He Praised in

287

1992," by Thomas B. Edsall, the *Washington Post,* December 16, 1998, page A02.

5. JEWISH CALVES AND CHRISTIAN LIONS

Page 135: "It wasn't until April 1994 that the ELCA formally repudiated . . ." From "Declaration of the Evangelical Lutheran Church in America to the Jewish Community," April 18, 1994, online at us-israel.org/jsource/anti-semitism/lutheran1.html.

Page 137: "Back in August 1980 the Reverend Bailey Smith . . ." From 1980 Religious Roundtable national affairs briefing in Dallas, TX, quoted in "Dabru Emet: A Jewish Statement About Christianity," online at religioustolerance.org/jud_chrr.htm.

Page 138: "In a resolution passed by their national convention . . ." From "Baptists Move on Two Fronts in New Effort to Convert Jews," by Gustav Niebuhr, the *New York Times,* June 14, 1996, page A12; "Baptists to Step Up Evangelism of Jews," by Christine Wicker, the *Dallas Morning News,* June 14, 1996, page 1A; and "Jewish Leader: Relations With Southern Baptists the Worst in Decades," by Richard N. Ostling, The Associated Press, September 30, 1999.

Page 140: "For example, in an October 1992 column, Pat Robertson . . ." "Pat Robertson's Perspective," October/November 1992, distributed to members of the 700 Club.

Page 141: "In the same vein Jerry Falwell . . ." Quoted in *The Religious Right: The Assault on Tolerance and Pluralism in America,* New York: The Anti-Defamation League, 1994, page 4.

Page 142: "In his 1991 book *The New World Order,* Robertson . . ." *The Religious Right*, page 24.

Page 146: "For example, in a sermon he delivered in January 1999 . . ." *Dallas Morning News*, January 20, 1999.

Page 146: "Dr. Bill Leonard, a Baptist . . ." from "Falwell's comments condemned as encouraging hate crimes against

Jews," by Jeffrey Weiss, the *Dallas Morning News,* January 23, 1999.

Page 147: "Speaking in February 2002 before the annual convention . . ." Quoted in "Bad Faith," by Peter Beinart, *New Republic,* March 25, 2002, online at www.tnr.com/doc.mhtml?I=20020325&s=trbo32502.

Page 148: "In April 2002, when Israel was under siege . . ." "We People of Faith Stand Firmly With Israel," by Ralph Reed, *Los Angeles Times,* April 21, 2002.

Page 151: "Perhaps the sage Woody Allen . . ." From "The Scrolls," by Woody Allen, in *Without Feathers,* New York, Random House, 1974.

Page 151: "For example, they describe candidates who support funding . . ." Quoted in "Right Wing Organizations: Christian Coalition," People for the American Way website, online at pfaw.org/pfaw/general/default.aspx?oid=4307.

Page 152: "On his *700 Club* television program . . ." Quoted in "Right Wing Organizations: Christian Coalition," People for the American Way Web site.

Page 152: "I got a very peculiar letter in reply." Details and quotations from letter by Pat Robertson addressed to author dated April 24, 2002.

Page 154: "The FRC has even stooped to using the war on terror . . ." Quoted in "Right Wing Organizations: Family Research Council," People for the American Way Web site, online at pfaw.org/pfaw/general/default.aspx?oid=4211.

Page 155: "As Jerry Falwell once remarked . . ." Quoted in "Back to School with the Religious Right," People for the American Way Web site, online at pfaw.org/pfaw/general/default.aspx?oid=4181.

Page 156: "For example, Joyce Meyer Ministries . . ." Details in next three paragraphs from "Stealth Evangelism: *Rage Against Destruction* Targets High Schools," October 16, 2002, online at adl.org/church-state/rad.asp.

Page 158: "For example, consider the infamous broadcast conversation . . ." From *The 700 Club,* September 14, 2001.

6. TROUBLED ALLIANCE

Page 165: "Some Jews were influenced by their growing affluence . . ." See, for example, "Jewish Liberalism Revisited," by Charles S. Liebman and Steven M. Cohen, *Commentary,* 102:5, November, 1996, page 51, and "Are American Jews Moving to the Right?" by Murray Friedman, *Commentary,* 109:4, April 1, 2000, page 50. Also see "The Strange Phenomenon of Black Anti-Semitism," by Larry Elder, April 24, 2002, online at worldnetdaily.com/news/article.asp?ARTICLE_ID=27378.

Page 166: "At its very inception, the Black-Jewish rift was analyzed . . ." *The Anatomy of Frustration: An address delivered by Bayard Rustin, director of the A. Philip Randolph Institute, at the 55th National Commission Meeting of the Anti-Defamation League of B'nai B'rith,* Anti-Defamation League, New York, 1968.

Page 170: "Ironically, a striking illustration of this thesis recently emerged . . ." From "Surprisingly, Supreme Court Justice Clarence Thomas Shares View on Cross-Burning," by Stephanie A. Crockett, BET News, online at bet.com/articles/0,,c1sc77gb4874–5594,00.html.

Page 174: "In Georgia, for example, the local system . . ." "Don't Blame the Jews for Cynthia McKinney's Defeat," by Stephen Zunes, December 27, 2002, online at www.common dreams.org.

Page 179: "Leonard Jeffries, the former head of the Black Studies Department . . ." Details in the next three paragraphs adapted from "Anti-Semitism and Black Student Groups," online at www.adl.org.

Page 180: "The Recording Industry Association of America reports . . ." "One positive note in pop-music blues," by Steve

Morse, Boston *Globe*, December 29, 2002, online at www.ae.
boston.com.

Page 180: "But too many popular Black hip-hop artists . . ."
Details on Public Enemy from "The Rap on Chuck D," by
Andrew Wallenstein, online at www.generationj.com.

Page 182: "Scalded by widespread criticism, including a public
rebuke . . ." Quotation from letter by Michael Jackson
addressed to the author, dated June 22, 1995.

Page 183: "Founded in 1930 and led for over four decades . . ."
"The Nation of Islam," by Franklin Foer, October 19, 2001,
online at slate.msn.com/id/1075/.

Page 184: "Since 1975 the NOI has been led by Louis Farrakhan."
Biographical details from NOI Web site, online at
www.noi.org/mlf-bio.html.

Page 186: "Soon after his return from this tour . . ." Quotation
from "Minister Louis Farrakhan: In His Own Words," ADL
online at adl.org/special_reports/farrakhan_own_words/
farrakhan_own_words.asp.

Page 187: "His most well known speech (at New Jersey's Kean
College) . . ." Quotations from transcript of tape recording,
November 29, 1993.

Page 187: "He even tried to distance himself from his own
spokesman . . ." "Official Statement From the Honorable
Minister Louis Farrakhan and the Nation of Islam," February 17,
2001, online at www.noi.org/statements/bro_khalid02–17–
2001.htm.

Page 189: "Here is a sampling of public utterances by
Farrakhan . . ." All quotations can be found in "Minister Louis
Farrakhan: In His Own Words," online at adl.org/special_
reports/farrakhan_own_words/farrakhan_own_words.asp.
"From a speech in February 1998 . . ." Saviours' Day Speech,
Chicago, February 22, 1998. "From an interview in October

1998 . . ." Interview on *Meet the Press,* October 18, 1998. "From a speech in August 2000 . . ." Dallas *Observer*, August 10, 2000.

7. FROM HATRED TO *JIHAD*

Some of the information in this chapter has been adapted from the ADL report *Islamic Anti-Semitism in Historical Perspective* (2002), which may be found online at http://adl.org/ anti_semitism/arab/Arab_Anti-Semitism.pdf.

Page 199: "Theologian Muhammad Azzah Darwaza wrote . . ." Cited in *Arab Theologians on Jews and Israel: Extracts from the Proceedings of the Fourth Conference of the Academy of Islamic Research*, edited by D. F. Green, Editions de l'Avenir, Geneva, 1976, page 33.

Page 199: "Muhammad Sayyid Tantawi, a Muslim cleric . . ." Cited in "A Contemporary Construction of the Jews in the Qur'an," by Suha Taji-Faourki in *Muslim-Jewish Encounters: Intellectual Traditions and Modern Politics*, Harwood Academic Publishers, Amsterdam, 1998, pages 20–21.

Page 201: "In the same vein, Shaykh Abd-al-Halim Mahmud . . ." Cited in Yossef Bodansky, *Islamic Anti-Semitism as a Political Instrument,* The Freeman Center for Strategic Studies, Texas, 1999, page 84.

Page 201: "One participant at the Al-Azhar conference . . ." Cited in Moshe Ma'oz, *Current Anti-Jewishness,* page 39.

Page 201: "Similarly the Imam of the main mosque in Amman . . ." Cited in Bodansky, page 76.

Page 206: "In a children's supplement to the Brotherhood's *al-Da'wa'* publication . . ." Cited in Gilles Kepel, *Muslim Extremism in Egypt*, University of California Press, Berkeley, 1985, page 111.

Pages 208–209: Cartoons from various Middle East newspapers, as indicated.

Page 213: "Thus, 'The God of the Jews is not content' . . ." Cited in *Arab Attitudes To Israel* by Yehoshofat Harkabi, Transaction Publishers, New Jersey, 1974, page 273.

Page 213: "In August 1972 King Faisal of Saudi Arabia reported . . ." Efraim Karsh, "The Long Trail of Arab Anti-Semitism," online at www.aijac.org.au/review/200/263/essay/263.html.

Page 214: "Egypt's President Nasser endorsed the *Protocols* in 1958 . . ." Harkaby, page 235, and Bernard Lewis, *Semites and Anti-Semites: An Inquiry into Conflict and Prejudice,* W. W. Norton & Company, New York, 1986, page 208.

Page 215: "In December 1997 Mustafa Tlas . . ." BBC Summary of World Broadcasts, December 13, 1997.

Page 215: "Barely two weeks after September 11, 2001 . . ." BBC World Monitoring, September 24, 2001.

Page 216: "In one episode, three stereotypical Jews . . ." From "'Protocols' Appear in plot as Egyptian TV Series Continues," ADL online at adl.org/special—reports/protocols—plot.asp.

Page 220: "For example, Iran has become a sanctuary . . ." ADL online at adl.org/holocaust/denial_ME/in_own_words.asp.

Page 226: Some of the material in this section is adapted from *Jihad Online: Islamic Terrorists and the Internet,* Anti-Defamation League, New York, 2002.

Page 236: "According to Professor Abdul Hadi Palazzi . . ." From "For Allah's Sake," by Abigail Radoszkowicz, the *Jerusalem Post,* February 14, 2001.

8. THE POISONED WELL

Page 246: "Exasperated over lack of progress in the Middle East . . ." "Bush vows to veto loan guarantees," by Allison Kaplan and Michal Yudelman, *Jerusalem Post,* September 13, 1991.

Page 247: "Take the case of Joerg Heider . . ." Adapted from "Backgrounder: Joerg Haider—The Rise of an Austrian Extreme Rightist," *ADL International Notes,* New York, October 1999.

Page 249: "But when, in 1994, an interviewer from *Vogue* . . ." *Vogue* magazine, January 1994.

Page 249: "Critic Neal Gabler wrote a fascinating history of the subject . . ." Neal Gabler, *An Empire of Their Own: How the Jews Invented Hollywood*, Crown Publishers, New York, 1988.

Page 250: "I'm happy to report that . . ." Quotation from letter by Dolly Parton addressed to the author, dated February 17, 1994.

Page 253: "In a similar vein, I think actress . . ." *Cooking in the Litchfield Hills*, The Pratt Center, New Milford, CT, 1993.

Page 264: "Most people who come to understand the scope of the problem . . ." Information in this section adapted from *Combating Extremism in Cyberspace: The Legal Issues Affecting Internet Hate Speech*, Anti-Defamation League, New York, 2000.

Page 271: "Take actor and director Mel Gibson." "Is the Pope Catholic . . . Enough?" by Christopher Noxon, the *New York Times Magazine*, March 9, 2003, pages 50–53.

EPILOGUE

Page 279: "There's no better way for me to close . . ." From Elie Wiesel's Nobel Prize Acceptance Speech, 1986.